# MILITARY VEHICLES OF WORLD WAR 2

2Δ
17E
DPX-699L

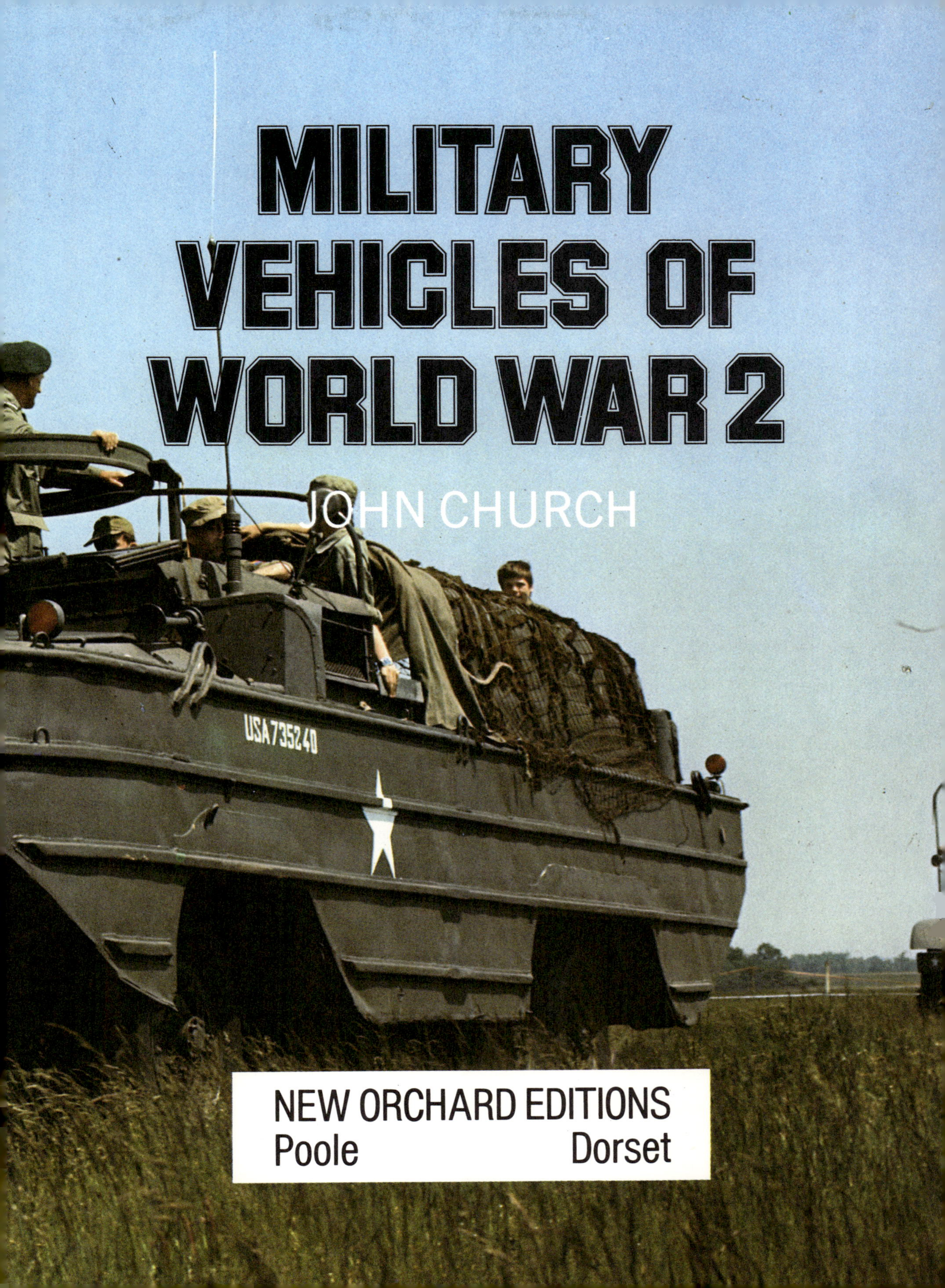


# MILITARY VEHICLES OF WORLD WAR 2

JOHN CHURCH

NEW ORCHARD EDITIONS
Poole Dorset

First published in the U.K. 1982 by
Blandford Press,
Link House, West Street,
Poole, Dorset, BH15 1LL



This edition published by
New Orchard Editions
Poole, Dorset

New Orchard Editions Ltd
Robert Rogers House,
New Orchard, Poole,
Dorset BH15 1LU, England

ISBN 1 85079 005 1



Typeset by Keyspools Ltd, Golborne, Lancs.

Printed in Singapore by
Toppan Printing Co. (S) Pte.

Colour photography by Andrew Morland

Title page *Intended for a limited life under operational conditions, the GMC DUKW-353 'Duck' confounded its early critics and remained in service for many years after WW2.*

# CONTENTS

# INTRODUCTION

In the eyes of the general public the 'Tank' and armoured fighting vehicles as a whole, together with their crews, have had an aura of glamour not accorded to the more mundane branches of the Army. It is satisfying to see that proper due is now being paid to the supporting arms without which the 'teeth' of any army would lack the essential bite to achieve victory.

Much has been written since the end of World War 2 about fighting vehicles and their development, and it is probably this wealth of information that has led the student of armoured warfare to realise that, without a vast and efficient support organisation supplying food, fuel and ammunition at the right time and place, providing efficient recovery and repair facilities, bridges, communications and medical attention, armoured formations are soon reduced to immobile masses of useless metal.

The first idea for providing mechanical self-propelled transport was the patent taken out in Britain by David Ramsey in 1634, but, like so many early inventions, its practical application was a failure due to the lack of a suitable source of power.

However, in 1765, James Watt applied steam to provide power and Cugnot produced the first steam-powered wagon in 1769. Although the Cugnot machine was unsuccessful as an artillery tractor the potential was there, and development of steam traction continued throughout the 19th century.

The Crimean War provided the first instance of the use of military mechanical transport when British heavy artillery was hauled into position by Boydell steam traction engines.

Later applications are to be found in the Franco–Prussian War and the Russo–Turkish War of 1878. The use of the Fowler engines by the Prussians jolted the French War Ministry out of its complacency and made it take a long hard look at the future of military transport. The British Army made good use of steam tractors during the Boer War, following their limited use in the Ashanti Wars of 1873, and the Aveling & Porter 'Steam Sapper' became the first steam traction engine designed especially for military use.

It was already apparent that the slow heavy steam tractors had a limited application with poor cross-country performance, and the idea of improving mobility over bad ground was crystallising into the application of endless track-laying machines. The germ of the idea seems to have sprung from Edgeworth, in 1770, followed by Thomas Germain's endless-track of 1801, with further variations on the basic theme by Palmer in 1812 and Barry nine years later, culminating in the all round track of Guillaume Fender in 1882. Applegarth then produced a tracked steam-powered tractor in 1886, followed by the Batter machine patented in the USA two years later.

These machines were heavy and slow and therefore unsuitable for the support of advancing fighting troops, being reminiscent of the cumbersome medieval siege trains. In fact, in Britain, the last applied operational use of steam traction was in providing the motive power for the 9.2 in. and 12 in. howitzers of the Royal Artillery Heavy Regiments.

The introduction of the internal combustion engine offering greater power-to-weight ratio, ease of operation, etc., rang the death knell of the military traction engine and opened up the possibility of providing true mobility for the armies of the world. In the years from 1904 to 1914 the internal combustion (IC) engine developed from its initial experimental stage into a reasonably reliable means of motive power. Thus, by 1910, the British Army had begun to recognise the potential of the IC engined lorry, and mechanical transport began to be adopted on a relatively large scale, although the motor-car had been used from 1902 for the transport of staff officers. In 1909, members of the Automobile Association moved a composite battalion of the Household Brigade from London to Hastings and back in their own private cars. In the previous year several Sheffield Simplex Standard 30 hp models had been used as gun tractors in the Grimsby area, travelling 80 miles at an average speed of 21 mph.

The value of the motor vehicle as an engine of war had, obviously, not gone unnoticed in other countries. In 1900, Lt Col Layriz of the German Army was advocating the use of motor-cars for the carrying of dispatches and general liaison work and by 1908 lorries were used on manoeuvres by the German Army to transport troops and supplies. The same year saw a mobilisation exercise carried out by the British Army in which London General Omnibus vehicles were hired to move troops from London to meet an 'invasion threat' in the Shoeburyness area.

An early advocate of the military application of the automobile was the French artillery officer Captain Genty, who operated a Panhard for a time in Casablanca in support of cavalry patrols. His use of an armed, but unarmoured, patrol vehicle in the desert over relatively long distances, foreshadowed the activities of the light car patrols in Palestine during World War 1 and the desert patrols during the North African campaigns of World War 2.

As befitted a nation at the forefront of automobile development, the French Army had purchased a variety of vehicles for trials at the turn of the century and operated one of the first 'subsidy' schemes – which are dealt with later.

The United States had trailed behind the European armies in the adoption of the motor vehicle, despite the efforts of Major Davidson in developing machine-gun armed cars. However, a couple of Packard 3 ton gun carriers were successfully converted to load carriers, leading to a small production order in 1910. This vehicle was the subject of several follow-on contracts and, with minor modifications, it remained in production until 1918, thus becoming one of the first standardised military transport vehicles.

The success of the petrol engined lorry in the role of cargo carrier led the military to develop other uses for the basic chassis and, apart from armed variants, specialist bodies began to appear in service. These ranged from pigeon lofts to ambulances, workshops and gun tractors.

The outbreak of war in 1914 led to an enormous increase in the demand for vehicles for use in support of the opposing armies.

Immediately the 'War to end all Wars' came to a close, demobilisation and disarmament were the order of the day. War surplus vehicles flooded on to the world markets, seriously affecting the development and production of new designs by truck manufacturers.

Severe financial restraints were placed on the military by those holding the government 'purse strings', but the lessons learned in the mud and carnage of France and Belgium were being carefully studied. The effect of the tank on military tactics was profound and it was obvious that the motor vehicle would have to play an increasingly important role in the support of mobile fighting units. There were, however, problems with the existing lorry in that equipped as it was with solid rubber tyres, its cross-country performance was inadequate. At the time, tracked supply vehicles were thought to be the solution.

In 1927, the British Army assembled an experimental mechanised force on Salisbury Plain, comprising a tank battalion, a tractor-drawn field artillery regiment, a lorry-borne machine-gun battalion, a Royal Engineers mechanised field company and supporting signals and supply units. Further trials were carried out in 1928 when radio communications for control of the 'battle' were introduced on a relatively large scale. Thus, the mid-1920s saw the beginnings of mechanised warfare; and the realisation that supporting vehicles were destined to play an increasingly important role was not lost on the armies of the world – particularly the Germans who had followed the manoeuvres of the British Experimental Force with keen interest.

The period from 1925 to 1935 was one of steady development, during which time the military, working within the strait-jacket of financial restraint throughout the time of the general trade depression, co-operated with the civilian truck manufacturers to produce vehicles suitable for their own national requirements. At the same time there was a constant interchange of ideas, foreign vehicles were purchased for evaluation and ideas incorporated into new designs.

These designs were beginning to come off the production lines in the late 1930s, as the war clouds gathered once more, and it was with these vehicles that combat was joined in 1939.

Overleaf *A Scammell Pioneer tank recovery/transporter of the early 1930s with a 20 ton semi-trailer showing the rear bogie removed for loading/unloading operations. The tank is a Vickers 16 tonner believed to be A6E2.*

# RESEARCH AND DEVELOPMENT

The term 'horseless carriage', used to describe the first motor-cars, was a very succinct description of the vehicle, as the frame, suspension and wheels were, in most cases, identical in design and construction to the original. In general, three types of frames and chassis were used:

*(A) Wood, re-inforced by steel flitch plates.*
This frame consisted of ash side members lined on one or both sides by steel plates, usually between $\frac{1}{8}$ and $\frac{1}{4}$ in. thick. As the flitch frame could be neither inswept at the front to accommodate the steering lock, nor upswept at the rear to allow for rear axle movement, it was out of favour by 1903 for car designs, but lingered on until the early 1920s for lorries.

*(B) Pressed steel frame.*
As soon as vehicle mass production reached the scale at which it was an economic proposition to install heavy presses, this type of frame came into its own. It had the following advantages:

*1* High strength to weight ratio.
*2* Cheap to produce in quantity.
*3* Allowed flexibility of design, in that it could be shaped to give extra strength or clearances etc. where desired.
*4* Ease of providing mounts for power plant, transmission body etc.

It was possible to design into the pressed steel chassis the desired amount of flexibility required for cross-country work, while still maintaining a rigid mount for the engine and gearbox, without having the excessive weight of a rolled channel chassis.

*(C) Rolled steel frame.*
This type of frame had limitations due to the rigidity of the rolled steel channel or 'I' sections forming the side members. Also, the construction tended to be heavier than actually required and, of course, in- and up- sweep were difficult to produce. This type of frame was gradually superseded by the pressed steel type. Thus, the majority of wheeled military transport vehicles at the outbreak of World War 2 had pressed steel frames.

These varied in detail design from manufacturer to manufacturer, those of Morris-Commercial having a triangular sub-frame

Top Left *An example of a German light military body on civilian chassis was the Mercedes-Benz 170V. Also used as a signals vehicle and for light repair work.*

Top Centre *Based on the pre-war Type 57, the Tatra 1.3/57K was produced in the early war years for the Wehrmacht as the Kfz 1. Later models had open-sided bodywork.*

Top Right *Close-up of left-hand front side of the Tatra showing digging spoon stowage and Notek blackout driving lamp.*

Left *The Austin 8 hp AP car 2-seater was a pre-war civil design adapted for military use and produced during the early war years for liaison work.*

*Thornycroft Type J chassis fitted with mobile workshop body. This chassis complied with Subsidy 'A' requirements and nearly 5,000 were produced from 1916 to 1918. Note the pressed-steel wheel discs with solid rubber tyres.*

carrying the engine gearbox and radiator mounted above the conventional frame.

It should be noted at this point that the British term 'rigid six-wheeler' did not refer to frame construction but distinguished a rigid from an articulated vehicle, e.g. two-axle tractor with a single-axle semi-trailer. This form of axle layout would now be referred to as a 6 × 4 or 6 × 6 as opposed to a 4 × 2–2 or 4 × 4–2.

The frame or chassis can be likened to the skeleton of an animal around which and to which the necessary appendages are attached.

Probably the most important of these appendages, and indeed the most difficult to get right from a design viewpoint, is the suspension. The strength of the springs has to be so proportioned to prevent the vertical movement of the axles being transmitted to the frame when moving over rough ground, and this problem is further compounded in the case of military cross-country vehicles by the vast load variations that can occur.

The initial lack of efficient shock absorbers added to these difficulties, as did the solid rubber tyre until the development of a successful pneumatic type.

Various spring arrangements have been tried. Prior to the introduction of coil springs, these consisted of variations on the basic multi-leaf spring, falling into the following categories:

The semi-elliptic.
The three-quarter elliptic.
The cantilever.
The quarter elliptic.
The transverse layout.

The most usual arrangement of springing was

*Dennis/Stevens searchlight truck of No. 1 Searchlight Bn Royal Engineers. Basically a Dennis Subsidy 'A' chassis with Stevens engine-driven generator giving 70 to 110 volt output. Forerunner of Tilling Stevens petrol-electric vehicles.*

the semi-elliptic for the two-axle vehicle, and various types of patented shackles were introduced during the late 1920s, one of the most interesting being the Albion spring-slide.

The Ford Motor Company was the chief exponent of the transverse spring system and applied it to both front and rear axles. On the Model T this allowed the axles to be as much as 15 to 20° out of parallel to the frame without imparting undue strain. This undoubtedly accounted for the success of the Model T over rough terrain during operations in the Middle East during World War 1. Radius rods had to be employed with transverse springs to locate the axle in the fore and aft phase.

Other manufacturers introduced interesting variations on the Ford theme. Scammell used a transverse spring on their Pioneer design first introduced in 1929, while FWD had a transverse spring connecting the rear end of the back axle semi-elliptic springs using special shackles. Garner used a similar idea on the front axle of their LSW 30 cwt 6 × 4 in 1929, while the Karrier W06/A 6 × 4 built for the Indian Army in 1928 had a conventional transverse front spring.

The German Army family of three-quarter tracked vehicles also featured transverse front springs. This was copied for the Bedford BT Traclat prototype gun tractor of 1944.

*Renault MH census no. T 17526. Extensively tested in Britain by the WVEE in the late 1920s following the Sahara crossing of 1923.*

The Humber Car Company used a system of double, quarter elliptic, transverse springs to give a form of independent front suspension (IFS) on their 4 × 4 8 cwt chassis on which were mounted the heavy utility ('Box'), the light reconnaissance car (Mk III), light ambulance and PU bodies.

Spring replacement and repair in the field has always presented problems to armies and can best be exemplified by the following episode that took place in the Western Desert at Tobruk in 1941–42.

No replacement springs were arriving from the UK, and the situation was critical as only a small amount of spring steel was available to the British forces. Salvage parties were therefore sent out to remove all springs from derelict British and Italian vehicles in the area, enabling 9 completed springs to be made daily by hand. As this did not meet requirements, heavy section Italian springs were reduced by drawing under a power hammer. Two 4 ft furnaces, two quenching tanks and forming jigs for the eye ends were manufactured from local resources. No temperature control equipment was available and temperatures were judged by the blacksmith. Four blacksmiths and four hammermen managed to raise production to 30 springs a day until Tobruk fell and the workshop passed into enemy hands.

The twin-axle rear bogie normally utilised two inverted semi-elliptic springs, attached above and below a central pivot bush and block bolted

*The original Pavesi prototype, without body, shows off its unique chassis articulation and method of steering and transmission.*

to the chassis.

Heavier duty bogies sometimes dispensed with the two springs, the axles being attached to a single large spring assembly.

The principle of the twin axle rear bogie is dealt with later.

Wheels originally followed the practice of horsedrawn vehicles, and were known in military circles as 'artillery-types'. The artillery-type was constructed by the wheelwright's traditional craft, with ash or hickory spokes built into a hub and a sectional rim of individual felloes bound by a shrink-on steel tyre. This type of wheel could be severely affected by climatic changes due to wood shrinkage etc. and was expensive to produce and maintain.

Wire wheels followed the principle of bicycle wheel spoked construction, and were much in favour on sports cars when built on to a Rudge-Whitworth hub, but never found favour with the military. A few examples were used on the light cars of the late 1920s and early 1930s, such as the Austin Seven and Morris 8.

Single-disc steel wheels were used on early cars, the disc was slightly dished for strength and a tyre rim riveted on to its circumference.

A heavier version used for lorries was made from a flat disc riveted to a dished disc.

This type of wheel was introduced by the British during World War 1 when supplies of cast steel from pre-war sources in Switzerland, Belgium and Germany were cut off.

With the introduction of the pneumatic tyre, consideration had to be given to the changing of covers as punctures were very frequent in the early days. The first idea was a detachable rim where the tyre and rim unbolted from the wheels and a spare tyre and rim slipped into place. Then the detachable flange was introduced to enable the heavier tyre to be removed without having to expend considerable effort in levering steel bead wires over the rims. This developed into the present WD type divided-disc wheel where the two halves of the wheel are bolted together. The retaining bolts are normally painted red to warn mechanics not to slacken the nuts before the

*Hathi artillery tractor under test in 1924 under difficult conditions. Note the overall chains fitted to all wheels.*

pressure has been let out of the tyre (the actual wheel retaining nuts being painted white).

Pressed steel disc wheels were introduced in the 1930s and continue virtually unchanged to the present day.

It was soon obvious that steel-tyred wheels were totally unsuitable for self-propelled vehicles and the solid rubber tyre was introduced. A rubber tyre was vulcanised on to a steel band and the assembly was then pressed on to the wheel-rim by means of a press requiring a load of 30 to 40 tons. Tyre dimensions were standardised at an early date and there were eight common sizes. Solid tyres remained in use for lorries into the early 1930s when the development of a heavy-duty pneumatic tyre at a reasonable price spelled the solid tyre's demise; artillery pieces and limbers continued to have solid tyres for several more years. When design requirements dictated that a dual rear wheel was required, it was normal practice to press two individual solid tyre rings on to a single broad wheel-rim.

Pneumatic tyres were developed for use on lighter motor-cars and cycles but the advantage of this tyre was obvious as it offered lighter chassis construction, better ride and handling and higher speeds. Also, from the military point of view, there was the question of enhanced cross-country performance.

Experience in World War 1 had shown that the solid-tyred 3 ton truck had very limited off-road performance and quickly bogged down in soft ground. However, the performance of the Packard and the Ford T light trucks in Iran and Palestine, together with the Fiat Model 15 TER in Italy convinced military authorities that a light (30 cwt load) pneumatic-tyred lorry was the answer. At this time the pneumatic tyre was in an early stage of development, small in section, expensive and prone to punctures and its application was limited to motor-cars and light lorries. The various tyre companies continued development and a significant step forward came in 1920, when a US Liberty Class B 3-tonner was converted by the Goodyear Tire Company into a rigid six-wheeler by the replacement of the rear

*The experimental 6 × 6 artillery tractor produced by the RASC from standard Hathi components and fitted with a WD-pattern rear bogie was probably Britain's first 6 × 6 military vehicle. Census no. T 17542, registration no. MH 3090.*

axle by a Goodyear-Templin single-axle four-wheel bogie with pneumatic tyres fitted to all wheels.

Until large section tyres became available in the late 1920s, the standard practice was to fit double wheels on rear axles. This was uneconomic as four extra wheels had to be fitted on a six-wheeler and, because of poor durability, two spares were usually carried, also the twin rear tyres suffered extra damage due to stones etc. wedging between the covers and damaging the side walls.

Progress in tyre design continued and new materials and techniques gradually overcame the early deficiencies.

Special treads and special purpose tyres became available in response to the military quest for better cross-country performance.

Britain, Italy and France gained a great deal of experience from motor-car expeditions across the deserts of North Africa in the 1920s and early 1930s. Probably the most famous being the Trans-Sahara crossing by the French Citroen-Kegresse team in 1922 and by Renault in 1923. Sand tyres were designed as a result. These were flexible walled, broad treaded, and imposed less pressure per square inch than a camel's foot, the most effective treads being shallow block treads or plain peripheral grooves. Sand tyres became standard fit for desert use on British Army vehicles from the mid 1930s, and proved their worth in the North African campaigns of World War 2.

The search for an effective cross-country tyre also continued and several companies produced characteristic tread patterns. Dunlop in Britain introduced 'Trakgrip', a non-directional cross-country tyre that became a common feature on pre-War army vehicles. The Americans introduced a similar type of non-directional tread but with grooves at right-angles to the tyre axis. Then came a 'directional' tread where the pattern consisted of deep 'V' chevrons. These tyres were classified as directional, because they had to be fitted with the point of the 'V' facing away from the direction of rotation in order to obtain maximum traction in soft conditions.

Cross-country tyres were expensive to produce and contained a larger amount of rubber than normal commercial tyres. Their use was, therefore, confined to certain classes of vehicle, and usually restricted to the driving wheels only.

A so-called 'bullet proof' tyre was developed just before the outbreak of war in 1939, and was known as the 'runflat' in Britain and as the 'Combat Tire' in the USA.

The general principle of the runflat design was that a stiff rubber inner-liner was cured to the inside of the sidewall. The stiffened sidewall was capable of supporting normal vehicle loads when deflated for distances of between 50 and 100 miles if speeds were limited to a maximum of 30 mph. Their use was limited to certain wheeled combat vehicles and artillery equipments but, if fitted to general service vehicles, then no spare wheel was carried.

In 1944 the British developed a tubeless runflat 7.00 × 18 tyre for use on Daimler scout cars. This was so successful that it formed the basis of a whole series of tubeless runflats for post-war British combat vehicles.

In order to prevent tyre creep on the wheel rims when running on deflated runflats, special bead-locks were fitted. Earlier pneumatics had bead-locks of different design fitted as a matter of course, but these were dispensed with as tyre/bead and wheel design progressed.

There was little standardisation of tyres during

World War 2 in the normal general-purpose or highway-pattern class, and the British Army had over 100 tread patterns in service and at least 50 differing sizes while the Canadian Army had 24 stock sizes.

Rubber became in short supply to both sides during the course of World War 2 and substitute synthetic materials were developed. Tyres made from butyl needed extra care and maintenance to achieve the life of a natural rubber tyre.

While the British and US Armies were evaluating the potential of the rigid six-wheeler for cross-country work in the immediate post World War 1 era, Adolphe Kegresse had returned to France from Russia following the Revolution bringing with him his design for a semi-track system. Kegresse had been in charge of the Czar's motor transport prior to 1917, and had perfected his semi-track design in 1910 to give his employer's vehicles a better over-snow performance.

The Kegresse system was adopted by Citroen and achieved early fame when Citroen-Kegresse semi-track vehicles were successful in making the first motor-car crossing of the Sahara in the winter of 1922–23.

The Kegresse track assembly depended for its functioning on the fact that the rear axle of the vehicle to which it was fitted carried no weight. The total weight was divided between the front axle of the vehicle and the load carrying assembly of the Kegresse suspension. The weight carrying axle had to be positioned as close to the vehicle centre of gravity as possible, and was located immediately above the two pairs of track rollers between the two suspension springs.

The driving pulley was in two parts and positioned at the rear end of the assembly, the outer half being driven by the vehicle rear axle. The inner half being driven by a cam-ring device which tended to close the two halves together, thus gripping the drive tongues of the track – the more power transmitted to the driving pulley the tighter the grip.

The front idler pulley was mounted on a cantilever arm attached to the weight carrying axle, and also contained the track tensioning mechanism.

The original method of driving the Kegresse track was not entirely satisfactory from the military viewpoint, so a 'positive drive' track was introduced. The basic layout was reversed, with a sprocket type driving wheel at the front transmitting power to the track via blocks bolted to the edge of the track. The track itself was kept centred by tongues in the middle that passed between the two halves of the sprocket.

*In the 1920s the British Army experimented with various types of semi-tracked vehicles, the 9th Regt Field Artillery being equipped with a selection for trials. The Citroen-Kegresse P7 battery car being one type.*

A competition was arranged in 1924 to demonstrate to the British Army the superiority of the Kegresse semi-track over conventional wheeled vehicles. During this demonstration, the semi-track defeated wheels over a treacherous piece of bog and the French representative of Kegresse threw his hat high in the air shouting, 'C'est la mort de la roue'. His hat joined the 'wheels' in the bog and the British Army purchased a trial batch of battery staff cars from Citroen-Kegresse but, as will be seen, they did not adopt the semi-track as a solution to obtain optimum cross-country performance.

The Kegresse system was, however, widely adopted by the French Army and apart from various Citroen vehicles was fitted to Somua and Unic tractors, many of which saw service in the German Army after 1940.

The Germans watched the Kegresse developments with great interest, and started their own design programme in the late 1920s. A standard design had been adopted by 1932, this used the Richter bogie system instead of the Kegresse design and a complete family of vehicles was projected to cover the following:

1 ton Sd Kfz 10 Demag
3 ton „ „ 11 Hanson-Lloyd
5 ton „ „ 6 Büssing-Nag
8 ton „ „ 7 Krauss-Maffei
12 ton „ „ 8 Daimler-Benz
18 ton „ „ 9 Famo

This weight classification refers to the towing capability of the vehicle and not its payload, the payloads were relatively small.

One automobile company was selected to act as design parent for each category (these are listed above alongside the type). Other companies could be used to boost production of particular models as required.

The Sd Kfz 2 semi-track motor cycle must be added to the above list. It was designed and produced by NSU and designated Kleines Kettenkraftrad (small tracked motor-cycle), but was commonly known to friend and foe as the 'Kettenrad'.

Production of the various types of semi-track vehicles for the German Army commenced in 1932/33 and continued until the end of the war in various guises. Captured vehicles were evaluated by the British and American experimental establishments, and by January 1945 the first British equivalent was being tested in prototype form.

Limited production of semi-tracks continued for a short while after the cessation of hostilities, some being issued to units of the British Liberation Army.

A feature of German design was track assisted steering. When tight turns were required, the controlled differential mechanism gradually came into play after the steering-wheel had turned more than half lock, speeding up the outer track and slowing down the inner.

The main drawback to the German vehicles was that they were extremely expensive to

*The Volkswagen 82. Popularly known as the 'Kubelwagen' it preceded the Jeep and over 52,000 were provided for the Wehrmacht. No TV series on WW2 should be without one!*

*An early trials vehicle was obtained from Mercedes-Benz and fitted with a test body.*

produce, each track pin had 80 needle rollers. They were also time consuming to maintain, as each track pin had to be individually greased at regular intervals, the Sd Kfz 7 had 110!

Interrogation by the British of captured German drivers of semi-track vehicles revealed that this greasing requirement was rarely adhered to on active service, and little detrimental effect seems to have been recorded.

Because of these drawbacks, the German High Command authorised, in 1942, the development and production of a revised track with dry pins, but few vehicles of the new series were produced.

In Russia the few roads were soon rendered useless by heavy military traffic, and the poor off-the-road performance of the Wehrmacht's 4 × 2 load carriers, coupled with a chronic shortage of 4 × 4 vehicles, forced the Germans to convert conventional wheeled trucks into semi-tracks.

This class of vehicle was generally known as the 'Maultier', and several designs of tracks were used, the most common being an adaption of the British Vickers light tank suspension.

Apart from load carrying tractors and recovery vehicles there were numerous armoured variants of the German semi-track series.

Although American semi-tracked agricultural tractors had been adapted for military use during World War 1, it was not until the early 1930s that serious interest was revived when a Citroen-Kegresse P17 was purchased by the US Army for test and revaluation.

Several firms produced 'half-track' conversions including Cunningham, Linn and Marmon-Herrington.

The Marmon-Herrington set a new standard in cross-country performance because the front axle as well as the rear track was driven.

This design was incorporated in the now famous family of standardised armoured half-tracks produced in vast numbers during World War 2. Several load carrying vehicles were converted by Marmon-Herrington prior to

*Guy FBAX of the Air Defence Experimental Establishment c 1930, fitted with special searchlight body. 'A' frame for offloading 90 cm searchlight and large cable reel fitted to rear of crew compartment. Generator driven from gearbox PTO (power-take-off).*

World War 2 to half-tracks with a driven front axle, the Ford $1\frac{1}{2}$ ton M2 half-track was a typical example.

British experience with the Citroen-Kegresse led to vehicles of local origin being converted to the Kegresse system, these included various models of the 30 cwt and 3 ton Crossley and Burford companies.

At about the same time, the British company, Roadless Traction Company Limited of Hounslow, produced their own track conversion for wheeled vehicles. This differed from the Kegresse design in that, instead of a continuous rubber and canvas track, the British track was made up of individual interlocking steel links. The bogie sprocket and rollers were rubber-tyred to reduce noise and afford some cushioning effect. Roadless Traction conversions were fitted to Guy, Morris Commercial and FWD (AEC) vehicles for the British Army and used in small numbers during the 1920s alongside the various Kegresse models.

The Roadless Traction Company later produced a tracked bogie known as the 'Orolo'. This was fitted to the British 45 ton tracked tank transporter trailer of World War 2, as well as to aircraft salvage trolleys of various sizes for the Royal Air Force. Orolo bogies were also fitted experimentally to the RASC experimental half-track, and to a Shelvoke & Drewry 30 ton semi-trailer tank transporter in place of the rear wheel bogies; the tractor being a Diamond T Model 981 with overall chains on the rear wheels.

Most of the British Army semi-tracked vehicles served with Royal Artillery regiments in various capacities alongside the fully-tracked Dragon tractors, until the Dragon tractors were replaced by the 6 × 4 tractors in the early 1930s. During World War 2 the British experimented

*Tilling-Stevens TS20 searchlight lorry. A much less complicated vehicle than the Guy. A 24 Kw generator driven by petrol engine provided motive power to wheels via electric motor, or to 90 cm searchlight.*

with Maultier type vehicles, substituting a tracked bogie assembly in place of the normal rear axle.

AEC converted a Matador using a track based on that fitted to the Valentine tank, while a Bedford QL and an Austin K5 had an installation based on the bren carrier design. None of these designs went into production.

Other countries experimented with semi-tracked military vehicles and Russia, adopted home of Kegresse, produced Kegresse tracked versions of the GAZ and ZIS trucks, some of which were in production throughout World War 2.

The success of the Kegresse system did not go unchallenged, and in 1923 Renault produced a six-wheel car developed from the standard 13.9 hp model. In December of that year, a special six-wheeler fitted with pneumatic tyres on twin wheels on each axle, made the Trans-Sahara crossing of 2,000 miles in five days.

The British War Office, impressed by the performance of the 'Desert Car', borrowed an example in 1924 for trials, and later purchased a chassis. This chassis was fitted with a gun tractor body and was subjected to intensive performance trials between 1924 and 1928 at the RASC MT School of Instruction at Aldershot. The suspension and tractive power of the Renault bogie were vastly improved by the RASC designers who produced a torque reaction system that kept the

*Morris 'Quad' prototype of 4 × 4 field artillery tractor with open body.*

wheels of both axles in ground contact with an even thrust. This eliminated the main fault in the Renault design, which was that the forward axle tended to lift at the critical moment when full drive was required from both axles.

In 1922, a small British force was sent to Chanak to prevent the Dardanelles being closed to international shipping by Mustapha Kemal. The RASC detachment was equipped with 30 cwt 4 × 2 lorries fitted with pneumatic tyres. This at a time when the cost and practicability of pneumatic tyres were viewed with trepidation by the commercial vehicle industry. However, the RASC vehicles acquitted themselves well and following successful trials with the Citroen six-wheeler, it was decided that future British Army load carriers would be pneumatically-tyred 6 × 4 vehicles with a WD type four-wheeled two-axle driven bogie.

By 1925, the medium designs by Karrier and Guy were being tested by the British War Department in severe comparative cross-country trials in competition with tracked, semi-tracked and prototype 4 × 4 tractors. In order to encourage other manufacturers to produce 6 × 4 designs, the 1923 subsidy scheme for 30 cwt 4 × 2 vehicles was revised in 1927 to include a 'medium', i.e. 3 ton 6 × 4 class; and further extended in 1928 to cover the 'Light WD Type 6 wheeler lorry', i.e. a 30 cwt 6 × 4 class.

Further efforts were made to improve the cross-country performance of the 6 × 4 vehicles, and the experimental branch evolved an overall chain for the rear bogie. This converted the vehicle into a semi-track and is a feature of British military six-wheelers to the present day, a tribute to its effectiveness.

The other development was the introduction of the limited slip, or lockable, differential. This device obviated an inherent drawback of the differential in soft ground, in that it prevented the majority of power being transmitted to the wheel encountering least resistance, resulting in wheel spin and loss of traction. In principle, all the differential lock did was to lock one axle-shaft to the differential cage. This action caused the whole gear-assembly to rotate, nullifying the differential action.

The basic 6 × 4 chassis layout became standard throughout the British Army and the Empire. The 3 ton (medium) version remaining in production throughout World War 2. The smaller 30 cwt light six-wheeler was probably more numerous during the 1930s, but quantity production had ceased by the end of 1940. Apart from load carrying general service (GS) or cargo bodies, these chassis were fitted with a wide variety of specialist bodies.

The cross-country performance of the 6 × 4 vehicle was quite good, if the load was kept within reasonable limits (3 ton reduced to $2\frac{1}{2}$ tons for cross-country), but the advocates of the four-wheel-drive (4 × 4) school were making their voices heard.

*Morris 'Quad' with revised body, the 'Beetle Back' style is beginning to emerge.*

The principles and advantages of four-wheel- or all-wheel-drive had been known in the early 1900s, but the production of load carrying 4 × 4 vehicles did not get under way until early in World War 1 with the introduction of the Duplex, Jeffrey (later Nash) Quad and FWD designs, many thousands of which were to see service with the Allies in Europe.

The US Ordnance Department then designed a 3 ton 4 × 4 truck incorporating the best features of the various US commercial products, and also bearing signs of influence by the French Latil designs. The Militor or Liberty truck was intended to become a standard type in the US Army inventory, but the Armistice was signed before large scale production could get under way.

The French and German manufacturers had produced 4 × 4 tractors early in World War 1, the best known being designed by Latil, Daimler and Ehrhardt.

An ingenious design originated with the Italian designer Pavesi. In his design, the chassis articulated both laterally and longitudinally; this, together with four-wheel-drive, produced a very good cross-country performance. The Pavesi P4 in later guises was in production well into World War 2 and served with both the Italian and German Armies.

Only the British neglected to develop a 4 × 4 vehicle during World War 1, and after the dust had settled on the conflict, the War Department began to evaluate, under the auspices of the RASC, samples of captured German vehicles including various tractors. A prototype 4 × 4 heavy artillery tractor was assembled from parts of the various German vehicles and christened the 'Hathi' (Hindustani for elephant). After testing and modification, the prototype was handed over to Thornycrofts who were tasked with 'Anglicising' the vehicle for production as the Hathi Mk 2; and 24 to 30 examples were delivered as artillery tractors, together with two or three, fitted with jibs, as breakdown vehicles.

The Hathi was a powerful vehicle fitted with a Thornycroft 11 litre 90 bhp engine, and the four-wheel-drive was permanently engaged. The front-wheel drive was through hollow stub axles and bevel gears to permit adequate movement for steering.

One example was fitted with a standard WD articulating twin-axle bogie in place of the normal rear axle, and fitted with dual pneumatic-tyred wheels on all three axles. This prototype took part in War Department trials in 1927, and had an outstanding cross-country performance when fitted with overall chains on the rear wheels. It was, incidentally, the first British 6 × 6 military vehicle.

The British War Department concentrated its efforts on producing a 4 × 4 field artillery tractor to replace the Dragons and 6 × 4 tractors of the 1920s and early 1930s, and the first 'Quads' appeared in service in 1938, followed by the AEC Matador medium artillery tractor. Thus the introduction of a 4 × 4 load carrier was delayed until February 1940, when the first Albion FT11 3 ton GS truck entered production, followed by the Crossley Q, Thornycroft Nubian, Austin K5, Karrier K6 and Bedford QL.

*Morris 'Quad' now with hard top and sloping rear to accommodate the 25-pdr traversing wheel, the holding-down attachments can just be seen.*

The problems of transmitting drive to a steering axle were gradually overcome, and culminated in the generally adopted Tracta design with constant velocity joints.

The majority of 4 × 4 designs incorporated a means of disconnecting the drive to the front axle when the vehicle was being driven on normal roads, this reduced front tyre wear and improved fuel consumption.

Gearboxes have evolved over the years from the early 1896 Panhard design, incorporating sliding gears with indirect drive on all speeds. This type of gearbox became known as the 'clash', or 'crash', because of the difficulty in meshing together gears running at different peripheral speeds without producing grating noise in the process. The driving technique of 'double-declutching' was the mystical art, by which drivers affected noiseless gear changes, before the days of syncromesh boxes. The author had the misfortune to be allotted a Guy Ant 15 cwt when learning to drive in 1943 at the D & M Wing at Bovington Camp. This was the only vehicle in the school with a full crash box, and to this day, when ever I muff a gearchange the ghostly voice of my instructor rings in my ears, 'Nine thousand nine hundred and ninety-nine revs too many'! I was not allowed to use the self-starter either, so stalling meant getting out and swinging the beast to start, and this led to rapid progress in mastering the finer points of handling the vehicle.

In 1904, direct drive was introduced on the highest gear, which meant the elimination of one pair of gear wheels, with attendant simplification of production and reduction in cost.

Early gearboxes had the gears mounted in line on a long shaft, this led to undesirable bending moments on the shaft. It also meant one had to change up or down through all the gears in turn. The gear change lever worked in a notched quadrant.

In 1905, the internals were re-arranged in two groups on separate short shafts operated by a change lever operating in the now familiar 'gate'.

In the period before the reliable synchromesh design was introduced, several types of epicyclic gearboxes were produced in an attempt to overcome the shortcomings of the crash box. The best known were the Lanchester and Wilson designs used extensively on British tracked vehicles. The Daimler preselector gearbox coupled to a fluid flywheel, was used only in the Daimler armoured and scout cars of World War 2.

Several attempts were made in the 1920s and 1930s to eliminate the gearbox and substitute a form of electrical transmission. This was generally known as 'the petrol-electric transmission', in which the normal vehicle petrol engine drove an electric generator. The generator provided power to either an electric motor attached to the back axle, or to motors on each of the rear wheels, thus there was no mechanical linkage between the power unit and the road wheels.

USA
363814
TP25

The British Tilling-Stevens Company produced many petrol-electric lorries and buses prior to World War 2, and the TS19 and TS20 searchlight lorries saw service from 1936 through to 1945, when many were snapped up by fairground operators to provide power for roundabouts and side shows.

There were also several attempts to combine mechanical transmission with secondary electrical power, as in the Pieper system and the Thames transmission. In the former, the number and weight of batteries required made the system impractical and the latter, with its combined generator and epicyclic drive, was not as efficient as the Tilling-Stevens system.

Other countries toyed with petrol-electric vehicles, chiefly in either the searchlight or workshop roles. Early examples from France being the Krieger and Aries searchlight lorries prior to World War 1.

The number of gears and ratios varied according to the basic design of the vehicle. A wider range of ratios was often obtained by fitting a two speed or High/Low auxiliary gearbox, or a two speed rear axle, a familiar design being the Eaton axle. Some designs, such as FWD (England), incorporated an epicyclic reduction gear in the wheel hubs.

The clutch evolved from the early leather cone design with either external or internal cones, via the single plate, to the multiple disc designs using either cork or Ferodo-type friction surfaces. On military vehicles, this part of the vehicle was probably subjected to more verbal and physical abuse than any other part.

There has always been a wide variety of engines in military service, because in peacetime it is more economical to use commercially available power units; while in wartime, quantity production is often more important than standardisation. Some degree of standardisation can, however, be achieved within an individual manufacturer's range of products. A standard series of engines was introduced in Britain after

*The Jeep – the Willys MB truck ¼ ton 4 × 4. Probably the best known military vehicle of all time and progenitor of many post-war field-car designs, not all of which equalled the original.*

*The first prototype Albion CX33 8 × 8 tank transporter tractor incorporated four wheel steering on the front and rear axles. The second prototype reverted to front axle steering and 8 × 6 drive.*

World War 2. This was the Rolls Royce 'B' Series of 4, 6 and 8 cylinders-in-line engines using common cylinders etc.

The variation within engine design is too vast a subject to be dealt with in one book. However, as with many inventions, the main progress has been in improved materials allowing refinements in design and higher performance/efficiency to be obtained. For instance, although engine speeds increased 100% between 1906 and 1914, valve-gear mechanical problems were not really overcome until the rapid development of aero-engines during 1914–20 led to the introduction of new alloys and improved manufacturing techniques. The majority of engines were either 4, 6 or 8 cylinder in line or V-8 poppet valve, water-cooled designs. There were other variations on the basic theme, such as the sleeve valve or rotary valve designs. The Daimler Knight engine had two concentric sleeves, with ports instead of poppet valves, whereas the Argyll had only one sleeve. Lubrication was a major problem and sleeve valve engines were never used in great numbers in road vehicles, although the Bristol Aero Engine Company and Napiers produced a series of efficient sleeve valve aero-engines from the mid 1930s. The Darraque and Italian companies produced rotary valve engines, but neither were able to compete commercially with the poppet valve designs.

Two-stroke engines had not been sufficiently developed to appear in large numbers, and their use was confined almost entirely to motor-cycles.

Several types of air-cooled engines were fitted

*The light 6 wheeler class of GS truck came into service in the late 1920s. The last of the breed was the Morris-Commercial CDF 30 cwt 6 × 4, in production from 1933 to 1940.*

into military vehicles as it was considered that the elimination of the water-cooling system would be advantageous during desert warfare. The vehicle would also be less liable to being disabled by battle damage to the radiator, etc. As always, there were disadvantages. Air-cooled engines rely on either, the forward speed of the vehicle to provide a cooling flow around the cylinder fins or, in slower moving vehicles, a forced draught provided by a fan; the power required to drive the fan usually being calculated at 10% of the total output.

Also, the machining operations in cutting the cooling fins in cylinder barrels are expensive and time consuming, and fins tend to clog up in service, particularly in the desert where the sand sticks to any oil film from the inevitable leaks.

Armstrong Siddeley, having produced a successful range of air-cooled aero-engines during World War 1, turned their attention to vehicle applications and produced an air-cooled V-8, rated at 90 bhp, utilising many features of the earlier aero-engines.

This engine was designated the SC, as it was originally fitted in an experimental tracked supply carrier. It was also fitted to production series Dragon gun tractors, and a 4 cylinder development was installed in the licence built Armstrong Siddeley Pavesi 4 × 4 gun tractor of the late 1920s.

In Germany, Krupp AG produced a light 6 × 4 truck powered by a horizontally opposed air-cooled 4 cylinder engine developing 60 bhp, known as the 'Boxer'. This 1933 prototype became the Kfz 69 and was produced in large numbers from 1934 onwards.

Probably the most famous and successful air-cooled engine (apart from motor-cycle power plants), was the Volkswagen horizontally opposed unit, but so much has already been written about it that further comment seems superfluous.

Diesel engines have been produced by most industrialised nations for many years and have provided the motive power for medium and heavy trucks and buses. Their use in military vehicles has paralleled civilian applications, in that they have been installed in the heavier class (6 tons upwards) of load carrying trucks. In many cases these vehicles were basically civilian trucks adapted for military use. The main reason for the diesel engine being confined to the larger range of vehicles, was its size and weight. Small diesel engines were, however, successfully produced in Germany at this time.

It has often been said that the method of taxation on civilian vehicles inhibited engine development in Britain between the wars. The 'horse power tax' was based on the Treasury rating of the engine which was arrived at from the following formula:

$$HP = \frac{D^2 \times N}{2.5}$$

when D = Diameter of piston in inches

N = Number of cylinders

2.5 = Constant

As will be realised, this formula favoured small bore engines which had a long stroke to obtain sufficient swept volume. This tax was applied to private motor-cars and did not directly affect commercial vehicles, these being assessed on the unladen weight. What did happen was that manufacturers became small-engine orientated, and the largest engine in their private car range tended to be used as the basis for their commercial vehicle designs.

As an illustration of this point, consider three Bedford military trucks, the MW 15 cwt, the OX 30 cwt and the OY 3 ton. All were fitted with the same basic Bedford 6 cylinder 3.5 litre petrol engine, giving 72 bhp at 3,000 rpm. In the case of the MW, it was governed down to 2,425 rpm giving 22 bhp per ton. The OX engine was governed at 2,580 rpm and gave 16.45 bhp per ton, whilst the heaviest in the range, the OY was governed at 2,800 rpm, but only gave 11.2 bhp per ton.

*The prototype Morris CD/SW LAA tractor (census no. H 382455) with totally enclosed cab and gun crew compartment.*

The gear ratios were, of course, matched to the vehicle and were not identical. The heaviest vehicle did, however, tend to be underpowered, particularly for cross-country work.

This attitude, coupled with the £30 pa tax on commercial vehicles up to $2\frac{1}{2}$ tons unladen weight, and a speed limit of 30 mph, with a proportionally greater tax and 20 mph limit on heavier trucks, resulted in volume production being concentrated in the under $2\frac{1}{2}$ ton class of truck, which were under-engined for civilian let alone for military use. This shortcoming did not generally apply to other countries.

Due to the need to maintain production of existing types, few new engines were introduced during the war. When the need for more power arose, the tendency was to couple two existing engines together, to drive through either, a common gearbox or, a transfer box, as in the Garner Straussler G3 artillery tractor and the Terrapin 8 × 8 amphibian. In the former vehicle, one engine drove the front axle and the other the rear, but either could, in an emergency, drive both axles. The layout on the Terrapin was different in that the left-hand engine drove the left-hand wheels and the right-hand engine drove the right-hand wheels. Both vehicles

*The prototype Morris-Commercial 15 cwt GS truck of 1934. A winch was installed between the front seats and the low-slung radiator was protected by a shield.*

*This shot of the prototype Bedford MW 15 cwt 4 × 2 truck of 1937 should be compared with the production version, the bonnet slope being made pronounced.*

utilised two Ford V-8, 85 bhp engines.

The Bedford BT Traclat $\frac{3}{4}$ track field artillery tractor was powered by two Bedford 6 cylinder engines, giving a total of 136 bhp, coupled to a single gearbox.

Another experimental vehicle with a twin engine layout, was the Albion CX33 tractor 8 × 6 (originally 8 × 8). The CX33 had two Albion 140 bhp engines installed, one of which drove the front and rear axles, which incidentally also steered on the first prototype, while the other engine drove the third powered axle.

The Dennis Octolat originally started life as an experimental 8 × 8 (Octo) light artillery tractor, powered by twin Bedford 72 bhp engines, but was later modified to a 6 × 8 configuration with a Leyland 9.8 litre engine.

The Americans were fortunate in having engines commercially available, giving outputs of up to 275 bhp. They also had aircraft engines adapted for AFV use, rated at 450 bhp. These engines were installed in the various American experimental 8 and 12 ton cargo trucks and prime movers under development at the end of hostilities.

The two methods used for cooling engines have already been mentioned, but some further comments on the water-cooling system must be made.

The original system of circulating the hot water through a cooling radiator, relied on the thermo-syphon principle. The hot water from the cylinder-block rose to the radiator header tank and then descended through the cooling tubes as its temperature dropped. A large radiator area was required, and the header tank had to be considerably higher than the cylinder-block due to the slow movement of the circulating water. In military vehicles, even larger radiators and header tanks were required to cope with the higher temperatures generated in low gear cross-country work and desert operation. Also, the radiator on military vehicles was susceptible to damage from distortion. This was due to mountings-movement, as the chassis twisted when traversing uneven ground. Consequently, it was a requirement of the British War Department subsidy scheme between the wars, that radiators should be isolated as far as possible from shocks and stresses by incorporating a trunnion-mounting.

Thermo-syphon circulation was eventually replaced by the forced, or pump assisted, circulation system whereby a centrifugal pump, usually forming part of the cooling fan-spindle, was driven at approximately engine speed by a belt driven from the crankshaft pulley. The tall large radiators which were fitted to the British 6 × 4 3-ton vehicles were very typical of the period.

*Gas producer units were fitted experimentally to several types of British military vehicle. This Bedford OY installation is typical, with the unit mounted between the cab and shortened cargo body.*

The reader will realise that it was not practicable under operational conditions in freezing temperatures, for the cooling system to be drained whenever the vehicle was left idle for any length of time. Antifreeze additives were developed but considerable education of vehicle drivers was necessary at the beginning of World War 2, to avoid damage by neglecting normal anti-frost precautions. An example concerns an artillery unit that bivouacked near Evreux in France, during February 1940, and suffered 13 cases of cracked cylinder blocks overnight, including eight irreplaceable Scammell gun tractors. This was due to failure to observe instructions. In fairness, it should be recorded that this regiment later formed part of a division of the Eighth Army and during the advance from El Agheila of over 500 miles only eight divisional vehicles suffered mechanical breakdown. The lessons of proper maintenance had been learnt the hard way.

A worse case occurred during the winter of 1941/42 in Paiforce, on the supply route to Southern Russia, when approximately 600 vehicles suffered from cracked cylinder blocks, and many had to be simply abandoned and pushed over mountain passes to keep the route clear.

Ingress of dust and abrasive particles into the engine, via the induction system, has to be kept to a minimum if excessive wear of the moving parts is to be prevented. Various types of air filters have been evolved over the years, mainly of the oil bath or felt/paper element type. Because of the conditions under military operations in desert or dusty environments, the air filters had to be of much larger capacity to cope with particle extraction and still give a reasonable time between filter cleaning.

Despite the installation of efficient air filters, engine life between overhauls could be very low. For example, the average overhaul life of a British military motor-cycle engine was between

7,000 to 10,000 miles; when operating in the dusty conditions of Normandy, this life fell dramatically, to between 2,000 and 3,000 miles.

Engine lubricating oil too, had to be changed more frequently in dusty conditions in order to remove abrasive contaminants, all of which added to the maintenance and supply problems.

The number of types of oils and greases in use in the British Army at the outbreak of World War 2 was at a minimum, and consisted of an engine oil M 220, a compound oil C 600, a grease GS and straight mineral oils. Experience during the first winter of the war in France with the BEF showed that engine oil M 220 was too heavy for use in cold conditions and two new grades introduced, M 220 Special, and M 160 (for winter use). M 220 being retained for use in air-cooled engines only.

Grease GS, a lime based lubricant, was unequalled in quality; and although superseded in 1943, it remained in use because of its superiority, particularly when applied to vehicles prepared for wading.

The first of the Hypoid oils was introduced in 1941, and C 155, later known as Hypoid 80, was used for steering boxes, gearboxes, etc.

The introduction of American and Canadian vehicles into the British Army inventory led to an increase in the number of different lubricants required, particularly with regard to AFVs; and in 1942, the Paul-Pyron agreement was completed on the use of American standard lubricants in British vehicles.

One interesting outcome of the Paul-Pyron agreement was the introduction of detergent oils into British service.

Special instructions were issued for the changeover procedures, as it was thought that the flushing action of detergent oil might lead to the release of sludge within the engine, with consequent blocking of oilways and interruption of oil supply. There were some reports of premature engine failures, at the time attributed to the detergent oils, but these were never substantiated

*Between the wars the British WVEE obtained examples of foreign military vehicles for comparative trials. This example is a Tatra 6 × 6 tractor fitted with right-hand drive.*

Top *Another Czech product to undergo testing was a late example of the Tatra T82 6 × 4 command car fitted with a flat-four 2,500 cc air-cooled engine.*

Centre *Another Tatra car to undergo testing was a derivative of the T82. A feature of these cars was the mid-mounted spare wheels which could rotate if the car bottomed on rough going.*

Bottom *This Austro-Daimler-ADZK was another type to be tested at Farnborough immediately pre-war, and influenced the Morris Type Q designs.*

and the new oil was quickly accepted. The success of detergent oils in 'B' vehicles soon led to its use in AFV engines.

By 1945, the number of different types of lubricants had risen to such an extent that they were presenting an almost impossible logistics problem. A Joint Associated Services Lubricant Panel was set up to investigate and report on the question of relating British, American and Canadian grades. This panel succeeded in reducing the total number of grades from 361 to 118, including reducing types of grease from 86 to 18, 10 of the remaining 18 being for special AFV application.

Only one grade of petrol, grade III, was in use in the British Army for both 'A' and 'B' vehicles at the start of World War 2, while WD DERV was used for diesel powered vehicles.

Leaded fuels for AFV use were introduced with the delivery of American vehicles, mainly MT 75 (lead content 1.2 cc per Imperial gallon, also known as 75 octane) and MT 87. In 1942, MT 80 (lead content 3.6 cc per Imperial gallon) was approved for British AFV use and under the terms of the Paul-Pyron Agreement its use was extended to 'B' vehicles.

Almost immediately, there was a flood of reports from users in all theatres of war complaining of engine failures. The most common cause given being severe burning of valves. In the case of small generating engines, there was also a drastic reduction in the time between top overhaul. The valve problem was only overcome by the introduction of valves manufactured from an alloy steel containing 20% chromium instead of the 8% used previously.

The introduction of octane fuels resulted in severe maintenance problems as well as additional supply difficulties for the British Army from 1943 onwards.

The Germans were faced with a growing shortage of fuel as the war progressed, and the priority for high octane fuels was given to the Luftwaffe. Synthetic fuels of reasonable quality were available and use was made of 'producer gas' as an alternative to liquid fuels.

By 1943, the production of synthetic oil products in Germany reached 5.7 million tonnes per annum, representing 57% of their total fuel requirements.

The British tried out producer gas installations on Bedford OL and QL 3-ton GS vehicles. These were not introduced into service, but were retained as an insurance against a possible shortage of petrol due to the effects of submarine warfare. A major drawback for military use, was the size and weight of the gas production plant. It severely reduced the cargo space and load carrying capacity of the vehicle to which it was fitted. Civilian applications were, however, more widespread in both Germany and Britain, where trailer mounted plants were quite common.

*The Tatra V750 4 × 4 command car was evaluated in the late 1930s and took part in comparative trials including the mountain circuit in N Wales in 1937.*

*Many Japanese designs were based on American products. The Isuzu Type 94A however owed much to European light 6 wheelers and was widely used from 1932 onwards for many roles.*

# PRODUCTION

Prior to 1914, the British War Office operated a subsidy scheme whereby users who purchased a vehicle complying with the particular subsidy specification received a yearly payment. This scheme ensured that military vehicles were immediately available on mobilisation and, more important, already in production.

The subsidy or subvention scheme was in operation in most of the major European powers prior to World War 1 but it was not in operation in the USA.

The scheme continued in Britain after the cessation of hostilities in 1918 and was highly successful in ensuring that the latest technical developments were incorporated in military designs, such as the development of the pneumatic tyre for load carrying vehicles and improvements in cooling, braking and suspension systems.

This scheme was eventually abandoned in the early 1930s, when the financial stringencies of the slump, and the gradual digression of civilian requirements from those of the military, made the scheme less attractive.

In 1936, the British Government set up the Ministry of Supply to deal with the provision of military equipment required under the rearmament programme. Prior to this, the procurement of wheeled vehicles had been the responsibility of the Master-General of the Ordnance (MGO), except for those vehicles required by the Royal Army Service Corps which came under the department of the Quarter Master General.

Prior to the formation of the Ministry of Supply up to ten separate contracts were required and tenders invited for the provision of a

*The inter-war medium 6-wheeled lorry qualified for a grant under the WD subsidy if operated by a civilian company and the Crossley Medium was a typical example.*

*A Pavesi P-4 under test at the WVEE. Limited-licence production by Armstrong-Siddeley commenced in 1929.*

complete vehicle – chassis, body, cab, canvas for the tilt and making up into the finished items, down to separate contracts for the towing hooks and the spring drawbar to which they were attached!

In 1934, it became obvious that there was no hope of producing a standard military load carrying vehicle, as the limited financial resources of Britain were directed towards obtaining fighting equipment, and commercial production could provide a reasonable substitute. A War Office committee therefore advised that six-wheeler vehicles (the basic 3 ton 6 × 4 chassis of the period) should be provided for high performance duties, and standard 4 × 2 vehicles used as load carriers. As most of the high performance requirements were for workshop or specialised bridging lorries, it was decided that the bulk of these bodies would be retained in storage. The 6 × 4 chassis were fitted with general service (GS) bodies and were operated by the RASC, until general mobilisation required the change of body and re-issue to other formations.

There were two back-up sources in Britain from which vehicles could be obtained in an emergency. The first was peculiar to Britain, in that there were two private companies who held a considerable number of military-type vehicles on their inventory. These were hired out to the Army for the traditional Autumn manoeuvres and for the rest of the year they were used for forestry and other cross-country work. However, as the number of vehicles available from this source would not satisfy full mobilisation requirements, an 'impressment' scheme had been devised.

6
147

Top Left *The GAZ-67B. This was the Red Army's equivalent of the Jeep and remained in production from 1943 to 1953 seeing widespread service outside Russia.*

Bottom Left and Top Right *The Wehrmacht was unique in having a complete range of semi-tracked vehicles to standard designs, production being undertaken by several manufacturers. The Sd Kfz 10 semi-track light 1 ton was the first in the series and designed by Demag. Several variants with specialist bodies were produced.*

Legislation was introduced allowing suitable civilian vehicles to be drafted into the Army in a time of general mobilisation, and an Inspectorate of Supplementary Transport was set up to administer the scheme. Inspectors were appointed to areas of Britain; their task being to record and inspect every vehicle that was deemed to be of military potential and had a minimum of three years life available.

Owners of designated vehicles were told where to deliver their vehicles on mobilisation. The numbers originally involved totalled 14,000, but this rose to a total of 35,000 before the end of World War 2. The impressment scheme, although ensuring rapid availability of certain types of vehicles, had no influence on the design of military vehicles whatsoever.

It should be remembered that, although at the outbreak of World War 2 the British Army was the only completely mechanised army in the world, there were no specialised military transport vehicles in quantity production – the operative word being quantity. The only type being built in any numbers was the 15 cwt 4 × 2 infantry truck class.

Due to the considerations mentioned in the

*Leyland were renowned for their Terrier, Retriever and Hippo 6 × 4 vehicles. However, they did produce a 6 × 6 prototype, the 6WL-1, in the early 1930s.*

previous chapter, the position in the British commercial vehicle industry was that volume capacity was available only for types in the $2\frac{1}{2}$ ton unladen weight class.

To achieve rapid production expansion of types suitable for military use, existing civilian designs were modified to meet military specifications by improving ground clearances, fuel capacity, cooling, springs, the substitution of large tyres with cross-country treads and elimination of dual rear wheels particularly for desert use, together with the fitment of military pattern bodies, POW (petrol, oil, water) can carriers, tool and wheel chain lockers. This range of vehicles continued in production virtually unchanged throughout the war.

We have seen that the British concentrated on building 4 × 4 gun tractors prior to 1939 at the expense of load carriers. The users and operational staffs now demanded that such vehicles be made available. Therefore the various company design staffs started work on 4 × 4 designs, which were, at the insistence of the Ministry of Supply, based on standard components as far as possible, thus ensuring that approved designs could enter into volume production as quickly as possible utilising existing engines, gearboxes, clutches etc.

The evacuation of the BEF from Dunkirk in 1940, with the loss of nearly all its vehicles and other heavy equipment, meant that production of existing types could not be interrupted until the British Army had been re-equipped. Despite this setback, the first 4 × 4 load carriers were in volume production at the beginning of 1941 and Bedfords produced 52,245 QLs, while Austin produced 12,280 K5s and Thornycroft produced 5,000 Nubians to which must be added considerable quantities from Ford, Karrier and Albion, as well as Crossleys for the RAF.

The introduction of the 4 × 4 chassis, into factories geared to produce conventional 4 × 2 vehicles, was not without its problems. These factories were planned to build one engine, one clutch, one gearbox and one driving axle per vehicle, but the extra driven axle required for a 4 × 4, doubles the axle output requirement and machining facilities. To which must be added the extra gear cutting required for the transfer gearboxes and bevels. Also, the facilities for the quantity production of Tracta joints had only become available just before the outbreak of war and new production and heat treatment techniques were still in their infancy.

The change from 4 × 2 to 4 × 4 should, in theory, have resulted in a halving of the production rate. Yet despite the low priority for new machine-tools accorded to the motor industry, production of 4 × 4 load carriers in February 1944 was at the same rate as conventional 4 × 2 chassis. It was not possible for the switch to 4 × 4 production to be completed before the end of hostilities in 1945.

Another factor in the slow build up of production, was the fact that it took 30% more manpower to build a 4 × 4 than a 4 × 2. The motor industry had no labour priorities during the war, except in the heavy tractor class, and there was a constant loss of manpower due to conscription or transfer to work of higher priority, this resulted in a lower labour force in 1945 than at the end of 1942. It should also be borne in mind that the industry was producing a wide variety of other war material. Vauxhall were 'parents' for the Churchill tank, Leyland for the Cromwell series, while the Nuffield Organisation produced AFVs, Bofors guns, fuses, mines, torpedoes, aircraft and aero-engines as well as a wide range of vehicles. The design staffs were also kept busy on priority projects, so work on new vehicles had to take a back seat.

*This totally enclosed cab was designed by Carbodies Ltd but not adopted for Retriever production, although it may have influenced the Hippo Mk 2 design.*

Some efforts were made at standardisation and rationalisation of types in production. It was, however, only possible to achieve the measure of standardisation within a company's own range of products. For instance, a particular engine would become the standard power pack for a series of different sizes of vehicles within that company. Axles, clutches and gearboxes were also examples of this 'vertical' method of standardisation. The Americans tended to adopt a 'horizontal' pattern, in which a particular class of vehicle, e.g. a $2\frac{1}{2}$ ton 6 × 6 was built to a standard design.

There were six basic classes of chassis (other than gun tractors and tank transporters) in production at the outbreak of war. Once the requirements had stabilised after the withdrawal of the BEF, it was decided to drop the 8 cwt GS class as the Jeep and 15 cwt could cover this role. It was also the intention to drop the 30 cwt GS chassis to concentrate resources on 3 ton 4 × 2 and 4 × 4 production. In the event, the Morris 6 × 4 30-cwt remained in Austin's production until 1944. Thus, some measure of rationalisation was achieved, but the cry was always for more and more vehicles to equip the new divisions being formed.

At one stage, cessation of production of the 3-ton 6 × 4 chassis was contemplated and the 4 × 4 was intended to take over its role. But it was

*The chassis layout of a typical 10 ton 6 × 4 chassis is depicted in this shot of a Leyland Hippo Mk II.*

*An Armstrong-Saurer 5 ton 4 × 2 with diesel engine based on the parent Swiss company's Model 5BL was tested as a possible 4/5 ton GS lorry in the mid 1930s, but no production was undertaken.*

discovered that the specialist bodies designed to be mounted on the 6 × 4 could not be accommodated on the shorter 4 × 4 chassis without major re-design. Also the demand for 4 × 4 load carriers was so great at the time that none could be spared to replace the 6 × 4s, and this type continued in production until the end of the war.

The 15 cwt truck remained in production throughout the war, virtually unchanged since its inception in the mid 1930s. Its major modifications concerned crew comfort, with the addition of full width windscreens, cab doors and side windows. Although the 15 cwt had a reasonable cross-country performance, due to its having the same engine as the bigger vehicles from the same stable, it was decided that 4 × 4 versions should be introduced to supplement similar vehicles being delivered from Canada and the USA. It was not until January 1944 that the Guy Quad Ant 15 cwt 4 × 4 GS started to roll off the production lines to be followed in October by the Morris C8/GS. Both vehicles were based on the companies' field artillery tractor chassis, but were not fitted with a winch.

It was decided, in mid 1944, that enough steel could be released to produce GS bodies for 4 × 4 3 ton and 15 cwt load carriers, replacing the wooden-planked designs. Additionally, a number of Albion BY5 and Thornycroft Tartar 6 × 4 chassis destined to have Folding Boat Equipment bodies had a new design of all steel GS body fitted instead.

All-steel bodies had long been advocated by users and designers alike for the following reasons:

*(A) Serviceability.*
A steel body withstands wear and tear and is easier to repair under operational conditions. Welding facilities usually being available to deal with chassis repairs.

*(B) Resistance to deterioration.*
Although liable to corrosion if not properly protected, steel was totally resistant to fungi and

*The 3 ton 4 × 2 chassis was produced in large numbers throughout World War 2 and was basically civilian in design. The Leyland Lynx WDZ1 shown here with closed cab was a typical example.*

termites etc. Planners looking ahead to the Far East required steel bodies for jungle warfare as water absorption of wooden bodies eventually led to rot setting in.

*(C) Cleanliness.*
A wooden body could absorb oil, petrol etc. which could contaminate the food that might constitute the vehicle's next load. While a steel body could be easily sluiced down and cleaned out.

*(D) Weight.*
Although a 3 ton steel GS body could weigh up to 150 lb more than the equivalent wooden body, moisture absorption more than cancelled out this advantage in time.

# GERMAN PRODUCTION

The German rearmament programme started in earnest in 1933 and some extremely sophisticated vehicles began to make their appearance. Notable among these vehicles was the series of standard semi-tracked vehicles and later the Klöckner-Deutz 6 × 6 light truck and the Auto-Union/Horch 4 × 4 medium car series.

The appearance of these and similar vehicles at the pre-war Berlin Motor Shows was a portent of things to come.

The size of the German automobile industry in relation to that of the Wehrmacht meant that its potential for manufacturing military vehicles was not large. However, production from factories in annexed or occupied countries must be taken into account and vehicles from Skoda and Tatra in Czechoslovakia, Steyr from Austria and from the French automobile industry all appeared on the Wehrmacht inventory. As already stated, only Panzer divisions were fully mechanised and *in toto* only about 20% of the German Army was so equipped. In the later stages of the War, it was the general opinion that the German Army had as many vehicles as it could handle, but this situation was due chiefly to the lack of fuel.

The plethora of vehicles, both military and civil, in production in the 1930s caused concern to the military planners, and a drastic rationalisation programme was initiated in 1938 under the direction of General von Schell which inevitably became known as the 'Schell Programme'. The then current belief in German efficiency was somewhat shattered when this programme revealed that there were no less than 164 different types of dynamos, 112 brake cylinders, 113 starters and 264 lamp bulbs in production!

The production of vehicles was rationalised and the numbers of models in production were drastically reduced, from 52 to 19 for cars and from 113 to 30 for lorries. The military did not escape, and several of the more complicated standard series such as Einheits vehicles were dropped, while other models were simplified.

Manufacturers now had to produce such vehicles as directed by the military, and this often meant making approved types designed by other companies.

The load carriers were reduced to four general categories; $1\frac{1}{2}$ ton (1,524 kg), 3 ton (3,048 kg), $4\frac{1}{2}$ ton (4,572 kg) and $6\frac{1}{2}$ ton (6,604 kg) and in each class there was a Standard Type (S type) with 4 × 2 or 6 × 4 drive. The cross-country versions with all-wheel drive were designated A-type (Allradomtrieb).

The German designers did not adopt large section or balloon tyres as the British had done. They retained narrow-section dual rear tyres throughout the War on the majority of load carrying vehicles; and this lack of appreciation of the advantages of large, low pressure, tyres made itself apparent during the North African campaign.

Two factors may have influenced their reasoning, the first being synthetic rubber tyres run hotter than similar sized tyres made from natural rubber, the second being that major design work would have been needed on the vehicles, to even up the front and rear axle loads.

The supply and quality of synthetic rubber appeared to be adequate throughout the War and the techniques in production were extremely good. It is of interest that when the British produced the Bedford BT 'Traclat', an almost direct copy of the German semi-tracked tractor, they ran into considerable difficulty in their attempts to reproduce the track-pads.

The Germans made more use of air-cooled engines than the Allies and had three outstanding types in production, the Volkswagen 4 cylinder horizontally opposed 985 to 1,130cc, the Steyr V-8 and the Tatra V-12 14,825cc 220 bhp diesel unit. As in Britain, diesel power units were in the minority and were of the pre-war commercial heavy truck and bus category.

Another reason for the small number of diesels in service was that the German synthetic fuel refineries could produce petrol more easily than DERV. Also, the High Command were of the opinion that the petrol engine was more reliable and that in the event of capturing enemy fuel dumps, the stock was more likely to be petrol than diesel.

The German designers and production engineers were at a disadvantage due to the shortage, or complete lack, of such 'strategic materials' as nickel, chrome, copper and manganese. The effect of these shortages on German production turned out to be less drastic than the Allies had anticipated or hoped. Detailed examination and analysis of captured vehicles by the Allies revealed that considerable ingenuity had been shown in utilising lower grades of steel. The use of high manganese steel for track links seemed surprising, but one should remember that, due to lack of transporters, the German tracked vehicles probably had to cover more road miles on tracks than the Allied vehicles.

One area in which German technique lagged behind was the production of torsion bars; this despite the fact that very large numbers were produced for semi-tracked vehicles and AFVs.

The specification of the chrome vanadium steel used was found to be over 20 years old, and the torsion bars generally had a fatigue life of 44,000 cycles. This life was raised during investigations to 90,000 cycles by the relatively simple process of shot blasting, and over 400,000 cycles were obtained by pre-setting before installation.

The shortage of copper led to radiators being of very light design, this made them prone to damage from vibration etc. Bronze was also scarce, and bushes made from this alloy were replaced by plastic, which proved to be more reliable.

There were two items of British equipment found on the majority of German vehicles, one was the 'Autoklean' engine oil filter, the other was the 'Enots' One-Shot chassis lubrication system, patented by Benton and Stone. This system was operated by the driver pressing and releasing a plunger on the dashboard once or twice a day. This resulted in lubricating oil being slowly fed into various points of the chassis via capillary type tubing, and as many as 24 points could be serviced in this manner.

A hiatus arose in 1941 when all development and research on transport vehicles was stopped. Work was restarted towards the end of the War but the results were too late to influence events. In any case the majority of effort was put into defensive weapons including large AFVs. Twelve new engine designs were being developed when the War finished, these ranged in size up to a 16 cylinder 1,500 bhp power unit and included five two-stroke and seven air-cooled types.

# WATERPROOFING AND WADEPROOFING

Waterproofing of tanks had been undertaken in Britain in 1934 in conjunction with amphibious tank trials, and two years later further trials were carried out on the wading capabilities of light tanks. In 1937, waterproofed light tanks and wheeled vehicles waded ashore from LC(M)s in about 2 ft 9 in. of water at Start Bay, Portsmouth, and then returned to Bovington by road at the end of the exercise. Distinction had now been made between waterproofing for fording and for wading.

The requirement for fording was fairly elementary as the immersion time was generally short, the water usually fresh and fairly shallow. The term 'wading', however, covers the conditions under which a vehicle disembarks, from the ramp of a landing craft, into sea water and travels under its own power to the beach in up to 6 ft of water.

When the planners in Britain began to consider the ways and means of fighting back into occupied Europe, it was realised that the vehicles would have to be landed over beaches, as it was extremely unlikely that ports would be available in the initial assault phase. This assumption was confirmed by the Dieppe Raid.

Some groundwork on waterproofing had been

carried out by the MVEE at Farnborough and by the Department of Tank Design in 1940, and trials were started at the Commando and Infantry Assault Combined Training Centre at Inverary (previously known as 110 Force). Initial trials showed that, at that time, MT vehicles could not wade ashore from a landing craft, even if there was only 3 ft of water; and after many vehicles had been 'drowned', the War Office began to sit up and take notice. Accordingly, a detachment of technicians was drafted into Inverary to investigate suitable waterproofing materials and devise the necessary techniques to be applied to the individual types of vehicles.

The landings on Madagascar provided the first operational use of waterproofed vehicles and showed that the methods and materials employed required further development. This work was allocated to the DME branch of the War Office, and trials were undertaken at Largs, to formulate waterproofing instructions for the vehicles of the British First Army that would be used in the North African landings in November 1942. Some 36 different types of vehicles were covered, as well as those of the US 1st Armoured Division which landed at Casablanca.

The various UK manufacturers of 'B' vehicles were now incorporating waterproofing instructions in their maintenance manuals and the actual waterproofing was made a unit responsibility.

With the successful conclusion of the North African campaign, the Allies commenced preparations for the invasion of Sicily. By the time of the landings, in July 1943, over 15,000 vehicles had been waterproofed for the British invasion forces, this took over 150 tons of sealing compounds and a quarter of a million feet of rubber tubing. One of the more difficult tasks undertaken was that of sealing the Ford V-8 engine which powered 40% of the British forces load-carrying vehicles in the Middle East.

Left *This compressed air installation was fitted into Ford WOT-6 4 × 4 GS lorries which were known as 'D-D Compressor Lorries'. It is believed that they were used to re-charge wading-screen erection cylinders of Duplex Drive tanks, hence the D-D.*

5,000 vehicles were later waterproofed for the landings at Salerno and for the landings by the British 1st Infantry Division at Anzio.

Attention was then turned to the preparations for the invasion of Europe – Operation Overlord – and this proved to be a gigantic task. Several thousand new vehicles had already been waterproofed by the Ministry of Supply and placed in store when, early in 1943, it was decided to withdraw sample vehicles from storage and carry out a typical wade test. Every vehicle failed. The subsequent enquiry revealed that the sealing compounds had perished during the time the vehicles had been in store. Further investigation revealed that waterproofing could only be carried out between six to eight weeks before the actual wade. The experimental wading establishment at Instow became so overloaded with work that a second unit was formed at Weymouth in October 1943. These units carried out trials on the effectiveness of the manufacturers' waterproofing schemes, checked that the instructions could easily be understood by the crews who had to carry out the task, trained personnel and mounted large scale wading exercises from landing craft.

21 Army Group decided that 'B' vehicles should be capable of six minutes immersion in 4 ft of water plus 18 in. waves (AFVs had to cope with 6 ft plus 18 in. waves.)

Some 5,000 'B' vehicles were waterproofed for the assault divisions between February and June 1944 together with 14,000 follow-up 'B' vehicles and over 4,000 AFVs. The following figures show the effectiveness of the waterproofing methods and the way in which they were carried out.

Percentage loss of total vehicles driven between landing craft and beaches

*1* 1.5% launched into water deeper than specified.
*2* 1.3% bogged down or lost by enemy action.
*3* 0.15% lost due to inefficient waterproofing or bad driving techniques.

When considering these results, it should be borne in mind that while some vehicles landed in shallow water, a great many had a 300 yard wade from a depth of 6 ft. The results achieved were undoubtedly a major contribution to the success of the invasion.

Hardly had the techniques for wadeproofing vehicles for the invasion of Europe been perfected, than consideration was directed to the problems associated with wading in tropical waters. The initial requirement was for wading to a depth of 6 ft, plus 18 in. wave allowance for a period of six minutes, but further investigations for wadeproofing up to 7 ft were requested. The British Ministry of Supply had set up a Tropical Research Centre at Ibadan, Nigeria, and by the time hostilities against Japan ceased, in August 1945, wadeproofing schemes for nearly 350 various types and makes of vehicle had been conceived.

The problems encountered when wadeproofing 'B' vehicles were somewhat more involved than those for AFVs, as the hull of an AFV was fairly easily sealed whereas the chassis of a wheeled vehicle was open and totally immersed. Engines had to be effectively sealed against the ingress of water and sand etc., the ignition system insulated and the fuel system protected from contamination; at the same time, there had to be a continuous supply of air into the induction system, and an unobstructed exhaust which was free from back pressure. Any electrical leaks were immediately enhanced on immersion in sea water, and as batteries were normally mounted in exposed positions low on the chassis, they required special protection and venting. A vehicle had, therefore, to be in first class mechanical and electrical condition prior to wadeproofing, otherwise 'drowning' was inevitable.

It should also be remembered that the mechanical strain of driving a loaded vehicle off a landing craft ramp into 4 ft of cold rough seawater, over a shingle sea bed and up a beach was considerably greater than that experienced during normal cross-country operations.

Maximum tractive effort was needed just at the time when the vehicle's satisfactorily warmed up engine was plunged into cold water and all its breathers were sealed to force its blunt body through solid water. One can sympathise with drivers who were more apprehensive about plunging into the water than facing a hostile reception on the beaches. Adequate training, and the fact that individual crews waterproofed their own vehicles, helped to overcome this important psychological problem.

Apart from prime movers, there was a host of other equipment to be waterproofed and each presented its own peculiar problem. Trailers with box-type bodies, e.g. radar trailers, could be sealed perfectly, but they then became buoyant and consequently in danger of capsizing; this meant the depth of wade had to be strictly limited. Artillery pieces had to be partially dismantled, with vulnerable components sealed separately in waterproof bags. This led to extensive reassembly on the beaches. For example, the 3.7 in. HAA gun took an hour and thirty minutes to prepare for action once ashore.

A great deal of organisation was necessary prior to embarkation of waterproofed vehicles, because of the extremely limited mileage that a fully treated vehicle could run. Prior to 'D' Day, the British Army evolved a three stage plan. For 'B' vehicles this was carried out as follows:

*Stage 1* took place in the Concentration Area, situated up to 200 miles from the embarkation port, where final maintenance tasks were completed. Fixtures such as instrument panels were sealed. Anti-corrosion treatment was applied, and a dummy run was carried out on the fitting of induction extension pipes, etc. This operation took an average of 50 manhours.

*Stage 2* took place in the Marshalling Area, not more than 20 miles from the port. It was here that the induction extension pipes, etc., were fitted, and sealing completed except for the engine breathers. A further $4\frac{1}{2}$ manhours were needed for this task.

*Stage 3* saw the fitment of the engine breather seals and the radiator protection sheet, and the attachment of a towrope to the front of the vehicle for recovery purposes. The vehicle could only travel a $\frac{1}{2}$ mile in this fully prepared state, and the work was completed immediately prior to loading or actually on the landing craft. The final test came when the vehicle commenced its laborious run to the beach.

# AIRPORTABILITY

The majority of readers will, no doubt, be aware that the early successes of the German parachute and glider-borne troops led to the rapid formation of British and American airborne divisions.

However, the transport aircraft and gliders available to both sides presented considerable problems when heavy support equipment had to be flown in, due to their lack of carrying capacity and the problems of loading. The introduction of the Messerschmitt Gigant glider together with the later powered version and the British GA Hamilcar went some way to providing a heavy lift capacity.

By the end of 1943, the British War Department were considering the problems involved in moving a standard infantry division by air, either in the strategic reinforcement role, or tactically, to landing strips seized and secured by airborne troops. To distinguish this concept from the parachute and glider-borne divisions, such divisions were to be known as 'Airportable' and eventually the 52nd (Lowland) Infantry Division was earmarked for this role.

The Airborne Forces Development Centre had been formed in May 1943, at Amesbury Abbey in Wiltshire, and it was here that the specialised equipment of the British Airborne Divisions was tested and developed, and loading techniques were evolved.

*The Morris CD/SW 30 cwt light breakdown had a simple block & tackle hoist capable of lifting 1 ton. An example was used in early air-portability trials.*

*The Bedford QLW tipper was an airportable vehicle but had to be completely stripped down for airlifting in two Douglas C47 Dakota aircraft.*

It was natural, therefore, that the problems of making heavy equipment airportable should be handed over to this establishment to solve.

The Douglas C47 Dakota was the only transport aircraft available at the time for the airportable tactical role and, because it was basically a converted civil airliner, it had many drawbacks.

Loads had to be tailored to a maximum of 5,000 lb, while there were problems in length and width dictated by the size of the freight doors and cabin floor area.

These limitations meant that even relatively small vehicles such as the 15 cwt infantry truck had to be substantially modified to achieve airportability as it was impractical to split a 15 cwt into two aircraft loads. The Ford WOT-2 could be carried as a single load, but the body had to be cut in half and dismantled to negotiate the door.

The Morris-Commercial C8/GS 4 × 4 15 cwt and the CMP 4 × 4 15 cwt trucks were ruled out on account of their weight.

Development of airportable 15 cwt trucks was therefore set in motion, and in Britain the Austin K7 reached the prototype stage together with the Bedford 4 × 4 15 cwt APT (Airportable). In Canada, Chevrolet designed an airportable

*The Karrier K6 was another airportable 3 ton vehicle, but only if stripped down. The winch version was a much sought after vehicle within British units and saw useful post-war service.*

version of the 4 × 4 heavy utility and the Dodge D3/4 APT Model T 236 entered production. The former was very similar to the US built WC 52 weapons carrier except that it had the slightly larger Canadian 236.6 cu. in. engine (hence the designation T 236), and a 12 volt electrical system. It did, however, weigh 900 lb over the ideal top limit, and although accepted as airportable, its weight meant a reduction in the maximum range of the transporting aircraft.

In every case except the Dodge, the body had to be removed for loading, but at least it could remain in one piece!

The story was very different for the 30 cwt and 3 ton vehicles.

The Morris 30 cwt 6 × 4 breakdown was the subject of initial trials, as there was a need for a light recovery section in the airportable division.

If the rear axle was removed, the chassis could just be manoeuvred into a Dakota and the loose axle stowed separately. A second aircraft then carried the body, jib, welding and recovery equipment together with the six-man recovery team.

As a result of this trial, investigations turned to the 3 ton 4 × 4 GS class and the Karrier K6 (with winch), Bedford QL 3 ton GS with steel body and the QL Tipper APT were chosen for initial study.

The Karrier had to be almost completely dismantled, and formed two Dakota loads made up as follows:

*Load one*

The chassis with its axles, springs and wheels removed was fitted with four small handling castors. The axles and road springs and winch were included as part of the load.

*Load two*

This contained the GS body split transversely into two parts, the top part of the cab, the wheels and tyres and miscellaneous equipment.

Modification of the Bedford QL was more complicated as, apart from splitting the body in two just forward of the wheel arch, the rear 16 ins of the offside chassis-frame member had to be made detachable together with the rear spring hanger bracket, and two detachable panels had to be introduced in the right-hand cab and engine cowling lower corner. The vehicle could then be transported in two C-47 aircraft as follows:

*Load one*

The chassis, with engine, gearbox and lower part of cab in position, but with the springs and axles removed. This load included the petrol tanks and spare wheel carrier. The axles and separate springs. The propeller shafts and exhaust pipes, and the jacks.

*Load two*

This contained the body in two parts. The POW can carriers, tool boxes and skid chains. It also contained the upper portion of the cab, the doors and front wings, the road wheels and spare wheel, and the tilt and superstructure.

The Bedford QL 3 ton 4 × 4 tipping, with winch, (APT) was designed to be airportable within the C-47 limitations, it could also double as a GS lorry as its body was larger than the normal tipper. Like the QL, it still had to be dismantled into loads, but it was no longer necessary to remove parts of the chassis-frame as this had been shortened. The winch had to come out, to ease manhandling for loading and unloading.

An airportable version of the Ford WOT-6 4 × 4 3 ton GS was also developed along the lines of the Bedford QL, except that the cab had to be removed *in toto*.

The Austin K5 4 × 4 3 ton GS presented a

more difficult problem due to the width of its chassis-frame; this had to be cut immediately behind the cab and 'fish-plate' joints introduced. Complete dismantling of the vehicle was still necessary.

Airportable versions of the Canadian Military Pattern 4 × 4 3 tonners were also designed by Ford and Chevrolet.

A total of eleven different makes of 3 tonners were redesigned for carriage in the Dakota, and all had to be completely stripped down. The end of the war brought development to a halt and in the post-war years purpose-built transport aircraft, allowing drive-on/drive-off facilities, became available.

After attending a short airportability course, the author had nightmares about an airstrip where one half of the incoming aeroplanes carried Bedford chassis, while the other half carried Morris bodies; fortunately I always woke before being called upon to resolve the situation.

There were a number of interesting experimental vehicles produced for the Airborne Division, and numerous modifications introduced to make the Jeep more versatile in this role. In 1944, for example, Jeeps were dropped from Stirlings by parachute to the SAS behind the lines in France.

Similar airportability developments took place in the USA, with the C-47 as the standard transport aircraft. The Americans confined their experiments to the $\frac{3}{4}$ ton and $1\frac{1}{2}$ ton 4 × 4 classes of vehicles, and the same problems arose, i.e. the C-47 could not accommodate a complete vehicle.

# COLD WEATHER OPERATIONS

Although it used to be said that the sun never set on the British Empire, only in Canada were extremely low temperatures experienced, and it was natural that the development of motor vehicle operating techniques under sub-zero conditions should emanate from the Canadian Army's Directorate of Automotive Design.

With the exception of Russia, the Allies experience of sub-zero operations had been limited to the brief Norwegian campaign and the garrisoning of Iceland.

As quantities of transport vehicles were being supplied to Russia early in 1942, it was decided to ask the Russians for a report, and a suitable questionnaire was compiled and forwarded.

Information was particularly requested on starting problems at temperatures of −30°F (−35°C) and below. The Russians replied that they had few such problems, as they always lit a fire under their vehicles to warm them before attempting to start!

In 1942, the Canadians devised two standards of special equipment for cold weather operations. The first, Specification OA 199, was known as 'Winterised Equipment' and was intended for use in temperatures down to −20°F (−29°C) and the second, 'Arcticised Equipment', was intended for use down to −40°F (−40°C) under Specification OA 99. Only Chevrolet vehicles were manufactured to this specification.

The problems to be overcome were, generally speaking, associated with starting, and were caused by the high viscosity of normal lubricants and the rapid fall-off of battery output. There were also problems with the cooling system freezing, the crew heating and insulation, and the general lubrication.

All greases and other lubricants were to a low temperature specification and the oil in the differentials, transmission, winches, etc., was Hypoid 80 diluted with $12\frac{1}{2}\%$ kerosene. Similarly, the shock absorber fluid was diluted with 30% Stodarts Solvent and the brake system refilled with a Royal Canadian Air Force hydraulic fluid.

The engines were fitted with high temperature

thermostats, special neoprene electrical cable insulation and low temperature diaphragms in the fuel pumps to prevent cracking.

Induction manifold priming and engine oil dilution systems were also fitted, both installations owing much to aero-engine practice. In the former, 'High Test Gasolene' (volatile petrol) was drawn from a special one gallon tank under the cab right-hand side by means of a dash mounted priming pump, and at each stroke 9cc of petrol was injected into the intake manifold through fine spray nozzles.

Oil dilution was carried out at the end of each day's operation, with the object of thinning the engine lubricant to assist the next morning's cold start. This was achieved by drawing a measured quantity of petrol from the vehicle's normal fuel supply and injecting it into the crank-case oil. The Detroit Specialities Prediluter fitted to the Chevrolet vehicles, controlled the amount of petrol introduced into the crankcase, by a selector graduated in degrees Fahrenheit, which was set by the driver to the expected local temperature. Obviously, care had to be exercised in the use of oil dilution, as over enthusiastic application could soon lead to a breakdown of the lubricating properties of the sump oil.

In an attempt to improve crew comfort, a large capacity commercial vehicle fan-assisted cab-heater and defroster was fitted, and hot air was bled-off to heat the battery box. The cab interior was sprayed with flock to improve insulation and reduce condensation. The canvas side screens with 'Monsanto' transparencies were retained, but these often cracked in the very low temperatures, and were never very satisfactory.

An under-chassis heater was provided with each vehicle, together with a wrapround shroud for the lower part of the engine compartment and cab. The shroud was lashed into place to prevent wind blowing under the engine and gearbox, and the heater then suspended under the sump. The heater was a 'pot' type and burned petrol, fed from the vehicle fuel tank, at a rate of 7.5 cc per minute and produced approximately 14,000 BTUs per hour.

Vehicles fitted with heavy utility, ambulance and house-type bodies, were equipped with fan-assisted combustion heaters and extra thermal insulation was installed.

It should be remembered that an electrical storage battery loses performance as the temperature falls. Generally speaking, if a battery is capable of giving 100% discharge capacity at +80°F (+27°C) then at −40°F (−40°C) its efficiency is reduced to 10%.

A special 21 plate battery was produced to Specification OA 121 by Exide of Canada, this had as good a performance at −20°F (−29°C) as the standard 17 plate battery gave at 0°F (−18°C).

As the battery in the Canadian Chevrolets was situated in the cab, it was a relatively easy matter to introduce an insulated box coupled to the cab heater.

Finally, provision was made for all 4×4 vehicles to accept snow chains on all four wheels, and to achieve this, the clearance between the front wheels and wings had to be increased.

A further Arcticised Specification (OA 111) was introduced in 1943 for Canadian CMP Ford and Chevrolet vehicles supplied to the USSR.

In general, the modifications were similar to Specification OA 99, with minor variations between the two makes of vehicle. The oil dilution control dial was simplified and the instructions given in Russian. 'Superfix' combustion heaters were fitted for cab engine heating and a small cowled chimney protruded through the cab roof. Ford vehicles, however, had two heaters, one for each bank of cylinders, and therefore had two chimneys.

All house-type bodies were internally sprayed with a rubberised insulation compound, and extra wooden floors and heaters installed. Tarpaulins were manufactured from a special duck material that stayed flexible down to −40°F (−40°C).

Auxiliary engines for providing power for workshop lorries were equipped with oil dilution, special heaters etc., in a manner similar to the vehicle engine.

The Winterised Equipment Specification OA 199 was issued in 1943, both Ford and Chevrolet being built to this standard and identified on the vehicle data plate by (−20°F).

A simplified crankcase oil-diluter system was fitted, this consisted of a header tank fitted to the windscreen centre pillar, and was fed from the engine fuel-pump. When the tank was full, a three-way cock was changed over, shutting off the supply, and allowing the contents to drain into the crank-case. The dip stick had an asterisk marked above the FULL line, if the contents reached this point, it was an indication that no further dilution was required.

Transmission oils and greases were to the standard low temperature specifications.

Engine induction primers were fitted, together with high temperature thermostats, and voltage regulation was increased to a maximum of 7.8 volts.

Cab-heaters differed, in that the Ford vehicles had a hot air system deriving its supply from a heat exchanger on the engine exhaust, while Chevrolets had two hot water fan-assisted heaters for heating and defrosting.

Cabs were insulated with flock, and ambulance and house-type bodies insulated in a manner similar to Arcticised vehicles.

Provision was made for chains on all four wheels by improving front wheel clearance as in Specification OA 99.

Technical trials on the equipment ended in 1944 with Exercise 'Eskimo' in Canada; this tested certain vehicles and equipment under the dry cold conditions of a continental sub-Arctic winter. It also applied the lessons learnt from British, Canadian and US troops in Alaska, Iceland and Europe.

The trial was held at Prince Albert, Saskatchewan, and included an advance northwards with a subsequent return over bush, plain and muskeg (swamp with frozen surface).

Snow depth averaged 18 in. and the minimum temperature was −34°F (−37°C), although temperatures as low as −72°F (−58°C) had previously been recorded in this region.

The conclusions reached from the exercise were that the Arcticised modifications were satisfactory, and that vehicle casualties were no greater than those which occurred during normal operations. However, as was highlighted by wadeproofing, vehicles and equipment had to be maintained in first class condition if they were to withstand the rigours of a sub-zero environment.

Vehicles fitted with skis or runners were not effective for general use if the snow coverage was not 100%, and it was recommended that con-

*The Chevrolet C60L was typical of Canadian produced 3 ton load carriers. This is an 'Articised' model.*

sideration should be given to the development of a fully tracked snow vehicle.

Lubricants developed for use in sub-zero temperatures were satisfactory.

Finally, in passing, the Universal Carrier was condemned as being totally unsuitable for operation under the conditions likely to be encountered in winter warfare.

When the Germans invaded Russia, they had hoped and planned for a short campaign. But, as their lines of communication lengthened and the Russian winter set in, it became obvious that the majority of their transport was not up to the rigours of sub-Arctic operations. One of the few vehicles designed specifically for such operations, was the Steyr RSO/01 Raupenschlepper-Ost (Tracked Tractor-East). The 'Maultier' semi-tracked conversions of standard load carriers also proved their worth, together with the standard range of semi-tracked vehicles.

The relatively large proportion of air-cooled engines in the German inventory was to their advantage under Russian winter conditions.

The pressures on German industry for increased production of existing types and, as already mentioned, the virtual cessation of vehicle development in 1941, meant that special modifications to cope with conditions on the Eastern Front were slow to materialise.

# SPECIAL PURPOSE VEHICLES

In the early 1900s, the British War Office formed a 'Mechanical Transport Committee' to watch over the development of the motor vehicle and its introduction into military service. This committee laid the foundations for future experimental establishments and service acceptance trials by undertaking various design studies and tests, including firing bullets into a can of petrol. They also commented on the unworkable complications of speed governors and complained about the lack of self starters. Many of the acceptance trials were carried out on the Brooklands Race Track which was conveniently situated close to the Army town of Aldershot.

In 1902 the responsibility for mechanical transport was vested in the Army Service Corps, on the recommendation of the MT Committee, and the Corps formed an 'Inspection Branch MT' which later came under the aegis of the MT School of Instruction and carried out experimental and development work in the early 1920s.

The MT Committee had now become the MT Advisory Board, but, in 1936, with the formation of the Ministry of Supply, it was re-styled the Mechanisation Board. At the same time, the Mechanical Warfare Experimental Establishment was formed at Farnborough, adjacent to the Royal Aircraft Establishment.

When the research and development of AFVs and transport vehicles were separated, the unit at Farnborough became known as the Wheeled Vehicle Experimental Establishment (WVEE) and also undertook trials on new auxiliary equipment, proof testing of tyres, etc.

Vehicles designed primarily for military use, were subjected to three basic series of trials:

*1* General Performance Tests.
*2* Special-to-Role Tests, e.g. gun tractors towing typical artillery pieces, tank transporters carrying maximum loads, etc.
*3* Reliability Tests.

Amphibians underwent preliminary flotation trials in the WVEE water tank, followed by initial water testing in Mytchett Lake near Farnborough, culminating in sea handling and beach landing checks at Instow.

Vehicles not intended for extensive cross-country work (known today as limited mobility vehicles) were put through a less arduous series of tests.

The area around Farnborough provided most of the differing types of 'going' required for cross-country tests, and man-made surfaces supplemented the natural terrain.

Initially, performance tests would be carried out on what was known as the Hindhead Circuit. This consisted of approximately 26½ miles of typical English country roads, never very wide and with numerous junctions, added to which were 9 miles of twisting gradients up to 1 in 9. The Salisbury Run then gave the opportunity for high speeds along the main road, and the Mountain Circuit in South Wales provided steep gradients of up to 1 in 6, with hairpin bends, and long pulls of 3 miles at 1 in 10.

Within the WVEE Test Ground, there were test gradients ranging from 1 in 4 to 1 in 2.25 on concrete surfaces and from 1 in 5.1 to 1 in 2.43 on gravel. There were also concrete slopes to simulate the angles of approach and departure from landing craft ramps.

Ground clearances and axle/bogie articulation could be measured against design figures, on a series of specially constructed concrete blocks.

Prior to World War 2, an area known as Hungry Hill had been inset with granite blocks to represent conditions extant on the North-West Frontier of India. However, the ravages of time had made this test area less effective, and when the North African campaign revealed weaknesses in chassis and suspension design, a new test section was constructed. This corrugated course consisted of concrete ribs set in gravel at irregular intervals and protruding about 2 in., inter-spaced by channels of varying width and up to 12 in. deep. Cross-country vehicles carried out a 250 mile run at various speeds over this track.

Other types of terrain were available at Long Valley, which was also used as a tank training ground and was therefore either extremely dusty or muddy. Bracknell Brickfield provided various consistencies of clay, assisted by a resident water

Previous Pages *The Scammell SV/25 heavy breakdown tractor was a popular recovery vehicle and it remained in service for many years. Despite no front-wheel drive it had an impressive cross-country performance, particularly when overall chains were fitted to the rear bogie.*

pump trailer. Ball Hill had a sandy loam texture and Fleet Pond was surrounded by sandy dunes. The whole area was covered with a mixture of grass, gorse and heather with pine copses.

A vehicle designed to a military specification would normally cover 10,000 miles on test. 7,500 on the road circuits and 2,500 miles cross-country, including 250 miles on the Hungry Hill and Corrugated Circuit.

During the trials, fuel and oil consumption were checked, speeds under various conditions and with various loads, including towed loads, were established, and braking efficiency recorded. Cooling efficiency was measured in the WVEE Dynamometer House and cross-checked during the subsequent running trials.

Turning circles were measured on a specially marked pan, and stability under the most unfavourable centre-of-gravity conditions checked on a special tilt bed.

Drawbar pulls on a variety of surfaces were checked against a dynamometer; in post-war years these were specially converted and instrumented Churchills and Conquerors. If a winch was standard equipment, then this too would be subject to a similar series of tests covering all of the roles, (side pulls etc.) and the maximum safe pull would be established, together with the efficiency of the winch limiting mechanism. Wading and waterproofing checks were made prior to sea trials at Instow.

On completion of the foregoing trials, each vehicle would be stripped down and the amount of wear on moving parts measured, alignments and adjustments checked. It would then be reassembled and repeat checks carried out to confirm figures originally recorded. Any major defects would be referred back to the manufacturer, for the introduction of suitable modifications, with comments on crew comfort and maintenance accessibility.

The vehicle would, upon successful completion of its trial, be released for service.

These trials were carried out on all new British vehicles and trailers, on lend-lease US types and on Canadian military pattern types. In addition, captured enemy vehicles were subjected to special trials at the WVEE, often supplemented by additional tests and investigations carried out by British automobile manufacturers under contract to the Ministry of Supply. The first enemy vehicle to be tested was a German aircraft refueller taken at Tromso in 1940 during the Norwegian campaign, and briefly used as a bulk fuel tanker, before being brought back to Britain after the evacuation of the Allied forces. Other German vehicles put through their paces at Farnborough included examples of the semi-tracked tractor family; these resulted in the Bedford BT prototypes.

*Trailer 7½ ton No. 1 Mk IV, Light Recovery. This was used throughout World War 2 for wheeled-vehicle and tracked-carrier recovery, usually towed by a 3 ton Gantry breakdown truck of the Unit light aid detachment.*

# RECOVERY

The problems associated with the recovery of battle damaged or mechanical casualties from the scene of conflict in mobile warfare, had been fully appreciated between the wars. It was realised that a considerable number of heavy recovery and load carrying vehicles capable of a degree of cross-country performance, equipped with winches and other specialist equipment would be required. However, the financial climate was such that little development work could be carried out, and procurement of specialist vehicle was on a small scale.

The British Army had the following types of recovery vehicles available at the outbreak of hostilities in 1939:

(*a*) Morris CD/SW 6 × 4 30 cwt breakdown. This vehicle was capable of light recovery work and had a winch with a $4\frac{1}{2}$ ton pull, but only had a jib, with a block and tackle capable of lifting 1 ton.

(*b*) 6 × 4 3 ton breakdown. This body could be fitted to various forward control 6 × 4 3 ton chassis and could cope with a suspended tow of $2\frac{1}{2}$ tons and had a 5 ton winch.

(*c*) Scammell 6 × 4 heavy breakdown tractor. This was fitted with a sliding jib capable of lifting a maximum of 3 tons and an 8 ton winch.

(*d*) Scammell 6 × 4-4 20 ton transporter. This semi-trailer combination was the only vehicle in service capable of handling the A9 and A10 Cruiser tanks and the Matilda I tank. It had a disadvantage in that the semi-trailer bogie had to be removed for loading and unloading, also, there were only two in service.

(*e*) A 4 wheel 5 ton recovery trailer. This was capable of transporting light tanks and carriers.

The Scammells were very good vehicles, but were in short supply as the same chassis was used for the heavy artillery tractor which had production priority throughout the war. Thus the Scammells were always in short supply, and had to be supplemented from US sources. An improved design of the 20 ton semi-trailer was introduced, this had hinged ramps which enabled loading to be accomplished without removal of the bogie. A 30 ton version later became available.

*Trailer 45 ton, Tracked, Recovery. Intended for battle field operations, towed by the Churchill ARV. The photograph is of an early production vehicle with unsprung bogies and unarmoured winch engine.*

In order to increase the number of recovery/transport vehicles available to the BEF, White-Ruxtall 922 and Mack EXBX 6 × 4 chassis were obtained from the US. These were shipped direct to France and fitted with British designed and built 18 ton capacity bodies. The majority of the BEF recovery equipment was lost and numbers were made up from White-Ruxtall 922, Mack EXBX and NR4 and White 920 vehicles; many of which had been ordered by the French as load carriers but not delivered and subsequently diverted to Britain. Immediately after the withdrawal from Dunkirk there was only a handful of 30 cwt and 3 ton breakdown vehicles distributed at workshops in Britain, whilst in the Western Desert there was one Scammell 20 ton transporter and several 3 ton breakdown vehicles.

In the aftermath of Dunkirk, orders were placed for a wide variety of breakdown and recovery vehicles (wreckers in US terminology). These included the existing British 3 ton 6 × 4 and Scammell 10 ton 6 × 4, supplemented by the CMP 3 ton range and the US Diamond 'T' 4 ton 6 × 6, the Ward La France 6 ton 6 × 6 series and the Mack LMSW 10 ton 6 × 4 vehicles. It was also realised that the 18 ton range of transporters could not cope with the next generation of AFVs, and specifications were issued resulting in the

*The Dodge WK-60 3 ton 6 × 4 breakdown, built to British requirements to supplement UK production. It incorporated right-hand drive and the rear bogie was driven via a Thornton power divider. British-built body was fitted to chassis in the UK.*

Cranes Mk 1 40 ton trailer and the US designed Rogers M9 45 ton trailer. The latter, in combination with the Diamond 'T' 980 tractor M20, was designated Tank Transporter M19 and was purchased in large numbers by the British even before the introduction of lend-lease. This combination was not entirely suitable for off-the-road recovery as the tractor lacked all-wheel-drive and the forward winching facilities provided on the 981; also, in loose ground, soil piled up under the relatively small multiple wheels of the trailer eventually stalling the tractor.

The shortage of Scammells continued to plague the recovery units and at the end of 1941 there were still only 29 vehicles in Britain, the best production promises being four per month.

About this time, the Director of Mechanical Maintenance (DMM) set up the Experimental Recovery Section to carry out research and development in recovery techniques and equipment, and to formulate specifications for consideration by the Ministry of Supply who would then proceed with production of approved equipment. Details of some of the products of the ERS will be given later in this chapter.

The intervention of the Germans in North Africa intensified the fighting, and an area of approximately 15,000 square miles of desert was strewn with the debris of war. The superior recovery techniques employed by the Afrika Korps received considerable attention from the British Press, though the War Office had been painfully aware of the situation for some time. The German ability to remove AFV casualties, particularly at night and even when retreating, was due to the availability of the 18 ton Sd Kfz 9 semi-tracked tractors and low loading trailers. The trailer was not, in itself, very extraordinary, and in many ways was inferior to the British and US counterparts. It weighed 13 tons, had a platform length of 18 ft 8 in. and was capable of carrying a load of 23 tons. Its ground clearance was only 18 in., and the rear bogie had to be removed for loading. Both front and rear bogies steered, and a steersman was employed on the rear bogie when on the move. Several of these trailers were captured and used by the British, who thought they were crude and over-complicated by German standards. Accurate alignment of the rear bogie was essential to ensure correct refitting, and in the heat of battle accurate alignment could be difficult! The rear steering was also reported to be decidedly skittish, particularly in wet or greasy conditions. The steering joints were, incidentally, lubricated by the British patented Enots one shot system.

Although, outside Tobruk, the British had to abandon a considerable number of AFVs which had suffered comparatively little damage, their technique improved rapidly despite continuing shortages of equipment. During the eleven day period of the Battle of Alamein, the REME recovered 530 tanks, of which 337 were returned to units of X Corps within that time. A typical return from Eighth Army, for the period 23 October to 20 November 1942, showed that 1,244 tanks and over 500 carriers and armoured cars were recovered, of which 1,007 were repaired. Recovery crews on both sides had to contend with enemy patrols intent on preventing the retrieval of valuable equipment; and on one occasion, a Scammell had its engine taken out of the chassis by a well placed 88 shot, the driver escaping shocked but unscathed.

The problem of removing disabled vehicles (particularly tracked AFVs) from the battlefield was the cause of much concern. By mid 1942, the British War Office Recovery Committee (of

*This shot of a CMP Chevrolet C60S 3 ton breakdown truck clearly shows the layout of the Holmes swinging booms and equipment lockers.*

*The Diamond 'T' Model 969 4 ton 6 × 6, with Holmes twin boom lifting gear, saw widespread service post-war with civilian garages.*

which the ERS was an offshoot) had reached the conclusion that the carriage of disabled tanks over battle grounds could not be resolved in the short term.

The policy then changed to developing an armoured recovery vehicle that could operate over similar terrain to the fighting vehicles, and could tow casualties to a point where they could be loaded on to a conventional recovery transporter and removed to the appropriate workshop facility.

In order to fill the gap before armoured recovery vehicles became available, consideration was given to modifying the Caterpillar D8 tractor for recovery purposes. This tractor was in widespread use in the USA and was being supplied in quantity to the British Royal Engineers as a bulldozer. A number of D8s were diverted for use by the REME as recovery tractors and proved quite capable of unditching the heaviest AFVs in service. There were, however, a number of disadvantages, but these were quickly overcome. The D8s were very slow, having a maximum speed of 4.9 mph and were unsuitable for prolonged road journeys on their own tracks. In British service, the D8 was usually moved on a 20 ton multiwheel low-loading trailer towed by the Recovery Section Scammell breakdown lorry. It had to be accompanied by a Loyd carrier which acted as a tender, as the D8 had no provision for the stowage of recovery gear. The Loyd was packed with gun planks, wire ropes, skid pans, earth anchors, jacks, snatch blocks and small tools needed to effect a recovery. The last drawback to the D8 was that the winch pull was limited by the ground resistance of the vehicle, this was overcome by fitting an earth spade designed by the ERS. This proved most effective and eventually became standard equipment on the majority of British heavy recovery vehicles. Use of the earth spade enabled the full 50 ton pull to be utilised on these D8s fitted with worm drive winches, although care had to be taken to avoid snapping the winch ropes.

Similar tractors from Allis Chalmers and IHC were used, and about 50 D8s were fitted with an armoured waterproofed superstructure for use as Beach Armoured Recovery Vehicles (BARV) in support of the D-Day landings.

A $7\frac{1}{2}$ ton 6 wheel light recovery trailer was designed for use in conjunction with the 3 ton 6 × 4 breakdown lorry. Produced by Messrs Cranes Limited, this proved to be practically trouble-free in service. It was fitted with a hand winch and capable of transporting tracked vehicles in the Universal Carrier class and wheeled vehicles up to the weight limit.

The problem of recovering 'dead' tanks, i.e. those that could not be towed for a great distance but were still worth retrieval, was still under active consideration. Several designs of cross-country trailer were produced, but only one reached the production stage, and then only in a limited quantity. This was the 40 ton tracked recovery trailer also produced by Messrs Cranes. This trailer was mounted on four Roadless Traction 'Orolo' bogies of a type used by the RAF for aircraft crash recovery. It had a demountable winch compartment with an armoured 'cockpit' for the operator. The 8 ton Scammell vertical winch was driven by a Ford V-8 engine, and the trailer was intended to be towed by a Churchill ARV Mk 1 which did not have a winch fitted. The idea being that, with the introduction of the winch equipped Churchill ARV Mk 2, the trailer winch unit could be removed. The cross-country performance of this trailer was truly remarkable, and it could carry up to 70 tons, however, its maximum towing speed was 5 mph, and due to lack of springing it could not be towed over hard surfaces and the trailer had to be carried on a wheeled 40 ton trailer for long distance road movement.

Messrs Boulton and Paul produced 75 of these trailers during 1944/45 but they were never used in their intended recovery role. Instead, they were used to transport light railway locomotives in LSTs from Britain to Normandy, and then over the beaches to the nearest railway track, later on in the campaign, 79th Armoured Division utilised a few to move heavy bridging equipment.

Because of the continuing shortage of Scammell 30 ton recovery transport, attention was turned to improving the cross-country performance of the Diamond 'T' and Rogers trailer. The towed trailer was to be dispensed with and the Diamond 'T' modified to accept a suitable semi-trailer.

*The truck, heavy wrecker, 6 ton 6 × 6 M1; produced by Ward La France, as Model 1000, Series 2, was later upgraded to 10 ton capacity.*

USA 0072446

8574857
USA

Top and Left *The 'Schwimmwagen' KPz1/20 amphibious light car could only be used on slow and calm inland waterways. Its ingenious hinged propeller, connected to a 'Dog' on the rear axle differential, gave a max. water speed of 4 mph. Steering was by using the front wheels as rudders.*

Top Left *Diamond 'T' 969 truck 4 ton 6 × 6 wrecker. Introduced in 1941 with closed cab. The 969A open cab shown here entered production in 1943. Widely used by the Allies.*

Bottom Left *Autocar truck 4–5 ton 4 × 4 tractor (U-7144T). Powered by a Hercules RXC 112 bhp petrol engine and used as the prime mover for a large variety of US Army 10/11 ton semi-trailers.*

A 30 ton semi-trailer was designed and produced by Shelvoke and Drewry, and a suitable fifth-wheel coupling fitted to the Diamond 'T' after the removal of the ballast body. By December 1942, the ERS were undertaking comparative trials between the Diamond 'T' conversion and the standard Scammell. The cross-country performance of the Diamond 'T' conversion was equal to that of its predecessor, with the added advantage of higher road speed. The Ministry of Supply ordered 200 Diamond 'T' conversions for recovery work, and the remaining Scammells, with the old 20 ton semi-trailers, were converted by slightly modifying the fifth-wheel coupling to accept the 30 ton Shelvoke and Drewry semi-trailer. In an attempt to improve the cross-country and desert performance of the Diamond T/S & D combination, the four-wheel trailer bogie was replaced by a Roadless Traction Orolo unit, and overall chains were fitted to the tractor bogie. Although trials results were satisfactory, the production problems and poor road performance of the Orolo bogies put an end to further development.

The recovery of wheeled vehicles did not present the technical problems posed by tracked AFVs, but the British Army was always faced with a shortage of suitable vehicles particularly in the heavy class. The Scammell was a very good vehicle, but numbers had to be made up with imports from the US and Canada where a wide range of suitable types was available. The Diamond 'T' Model 969 4 ton 6 × 6, in the medium class, and the Ward La France Model 1000 series 6 ton (later rated 10 ton) 6 × 6, in the heavy class, were very popular, although neither possessed all the features of the ideal recovery vehicle. The ERS was therefore tasked with producing a specification for a medium recovery vehicle embodying all the desirable features dictated by experience. A Bedford QLB light AA tractor had previously been modified by ERS to take a set of Holmes twin boom wrecking gear, but this did not achieve production status as the Diamond 'T' became available. In the search for a suitable chassis for the new vehicle, both the Diamond 'T' 6 × 6 and the FWD 4 × 4 chassis were examined, but the final choice rested on the Mack NM6 6 ton 6 × 6.

The final specification was issued in June 1944, and the prototype constructed, embodying Holmes twin boom lifting gear which could lift 6 tons when locked together or 3 tons when slewed outboard. It was also capable of a 3 ton suspended tow. The booms could be extended to give the high lift capability required when changing engines and all controls were operated from the driver's seat. A 7 ton winch incorporating a pendulum paying-on gear, developed by the British for the Diamond 'T' M20, was installed together with sprag-leg type earth spades capable of resisting a pull of 10 tons. The vehicle was designed to be airportable, and on satisfactory completion of acceptance trials in March 1945, production orders were placed for 500 vehicles destined for the Far East, but the cessation of hostilities led to cancellation of the contract.

Several types of ancillary recovery equipment were developed and produced by the ERS. These included the two models of Hollebone Drawbars for British and US AFVs, a draught connector to fit into the rear towing eyes of ARVs and a 25 ton snatch block for use with winches, enabling heavy recoveries to be made. Lightweight versions of standard equipment were developed for airborne and jungle operations.

# TANK TRANSPORTERS

The relatively short life of tank tracks when running on metalled roads had resulted in the development of a special class of vehicle capable of transporting tracked AFVs over long distances at a reasonable speed. During World War 1, because of the static nature of trench warfare, supply railways had been constructed close to the front line. It was possible therefore to deliver tanks by rail to the scene of operations. Transportation by rail did, however, have limitations. Quite apart from the fact that the railway system had to be in existence or constructed to serve particular locations, the size of the vehicle to be carried was governed by the loading gauge. The British gauge was less than that of the Continental or American systems, this had the side effect of influencing the design of British AFVs, imposing a restriction on overall width and therefore on the turret ring.

In order to achieve more mobility and less reliance on fixed rail networks, it was decided to provide heavy motor transport. The traction engine had provided the prime mover for shifting heavy and abnormal loads, but speeds and manoeuvrability were low and gave little advantage over a tank moving on its own tracks. The appearance of the Scammell Pioneer tractor and 20 ton semi-trailer in the late 1920s, provided an excellent transporter for the Vickers medium tanks of the period, despite the necessity to remove the trailer bogie for loading. However, the unit cost, combined with the financial restrictions that the depression placed on military budgets, meant there were never enough transporters to move more than a few of the small number of medium tanks in British Army service. Light tanks were less of a problem, as trailers that could be towed by standard medium lorries or breakdown vehicles were available in reasonable quantities.

Between the World Wars, other countries used heavy trucks to transport tanks in the $7\frac{1}{2}$ to 11 ton class. Most were 6 × 4 vehicles, usually equipped with winches, but some, like the French Berliet had an 'up and over' hoist. This system was an added complication which could only handle relatively light vehicles, and ramps were the normal method of loading.

In Germany, the Büssing-NAG 900 and Faun LD 900 transporters could carry a light tank such as the Pz II and also tow another on a low-loading trailer. Both were in service in reasonable numbers at the outbreak of war in 1939.

The US had few heavy military trucks in service capable of carrying a $7\frac{1}{2}$ ton light tank. There were, however, several heavy commercial vehicles available, and many were ordered by France and Britain to fill the 10 ton GS load carrier role in 1939.

Outstanding contracts for these vehicles were taken over by Britain after the fall of France, and the majority were then fitted with British designed and built tank transporter bodies. Vehicles so converted were, the Mack EXBX (the export version of the BX), the White 920 and the White-Ruxtall 922, all of which were rated at 18 tons and could cope with the Valentine tank. The early folding ramps operated by hand winches and known as 'Bees Knees', were replaced by a simplified one-piece design that could be manhandled into position. These converted vehicles were supplemented in 1940/41 by quantities of Mack NR4 6 × 4 13 tonners, the majority of which were delivered to the North African theatre of war for use in moving Stuart light tanks.

Later on in the War, when the weight of new tanks coming into service was beyond the capacity of the US produced rigid six-wheelers, the transporter bodies were removed and GS load carrier bodies refitted to those vehicles which had a useful life still left.

The Scammell 20 ton articulated transporter was in great demand and in short supply, and in 1941 Albion Motors introduced their CX24S 20 ton semi-trailer into production, however, problems with engine failures, together with other snags, led to the vehicle being derated to 15 tons. It then gave useful service as a GS vehicle, particularly when equipped with chocks etc., to enable large cable drums to be carried.

The shortage of transports in the 20 to 30 ton class in the British army, led the Purchasing Commission based in Washington to survey the US automobile industry in a search for suitable vehicles, and in 1941 they chose the Diamond 'T' Model 980 as a tractor to tow the Rogers 45 ton trailer. The Rogers trailer was the equivalent of the British Mk 1 and Mk 2 40 ton trailers. It was also produced by the Freuhauf Trailer Company, Pointer Willomette and the Winter-Weiss Company to meet British orders. The majority of these vehicles were delivered to the British and were classified as 'Limited Standard' by the US Ordnance Department, being designated 'Truck-Trailer 45 ton Tank Transporter M19', the tractor was designated 'Truck 12 ton 6 × 4 M20' and the trailer 'Trailer 45 ton 12 wheel M9'.

Vehicles delivered to the British, first saw operational use in the Western Desert where AEC Matadors were also pressed into service to tow surplus Rogers trailers.

Diamond 'T's were also used by British RASC tank transporter companies in the UK, and No. 534 Company was the first to be so equipped. It received its vehicles in April 1942 and landed in Algiers in February 1943 as part of the First Army, to be joined by 228 Company, each with 65 tractor/trailer units.

During the Tunisian campaign, each transporter averaged 9,850 miles. The total mileage covered was 618,464 miles during which 1,666 tanks and SP guns were carried. The average journey was 165 miles, but there were some long hauls of 1,500 miles.

Although the cross-country performance of the Diamond 'T' precluded it from off-road recovery work, it was available in vast quantities and formed the majority of the 9,000 transporters

*This trailer was designed as a transporter for medium tanks in the 1920s. The bed of the trailer was lowered by screw jacks and detached from both axles for loading.*

available to the British 21st Army Group in NW Europe.

However, the Diamond 'T' was not without mechanical problems. It suffered from cracked cylinder blocks, cracking between valve seats, failure of injector shafts and weak radiator mounts. The trailer rear bogie trunnion-bushes often seized, due to inadequate lubrication of inaccessible nipples; this led to the failure of welds due to flexing and resulted in loss of the complete bogie.

After August 1943, the closed cab was replaced by an open design with a canvas hood and an M36 mount for an anti-aircraft machine-gun, although this was omitted from British deliveries. Early tractors had the winch controls in the cab for driver operation, but these were transferred to the left-hand side of the winch with a small platform for the operator. A problem with the winch, in fact the problem with most winches, was laying on the cable to the drum. This rarely laid back neatly coil after coil and often finished up with jamming/kinking of the wire rope. A pendulum paying-on gear was developed by the British Ministry of Supply in mid 1944 and fitted retrospectively to Diamond 'T' tractors. As the inventor installed a hand-operated winch on top of a tall ministry office block in London, and dangled the cable for some eight floors past various office windows, early tests were not without their moments of humour. As he laboriously wound up the cable, intent on checking pendulum weights and their effect on the laying-on process, his less seriously minded colleagues in the offices below busily attached a variety of cardboard fish to the hook.

One British modification to the Diamond 'T' was to raise the ballast body by 9 in. in order to provide clearance for overall chains on the rear bogie.

The Model 981 tractor was equipped with a forward winching facility and could be identified by cable guide-sheaves in the front bumper. This was used for self-recovery only, and not intended to retrieve AFVs, etc.

Mention has already been made of the conversion of tractors to accommodate a fifth-wheel coupling for the Shelvoke and Drewry semi-trailer. The parts necessary to carry out this conversion were supplied with each semi-trailer, and could be applied to any tractor after removing the ballast body, which could be refitted later by reversing the modification procedure.

Other extempore modifications were carried out during the NW Europe campaign when, due to the long supply lines from Normandy into Belgium, Holland and Eastern France, both the British and the Americans converted a number of Rogers trailers into load carriers by decking over the tank channels and fitting high sides made from Somerfield Track.

A further 65 trailers were modified by a REME workshop, enabling the trailers to transport mechanised landing craft from Antwerp to a Rhine crossing assembly-point north of Nijmegen. The final order for 40 trailers being received only five days before the actual operation.

The Diamond 'T' was perhaps one of the best known American vehicles of World War 2 and remained in British service into the early 1970s, having been re-engined with a Rolls Royce C6N diesel engine.

The Federal 604 tractor with 20 ton semi-trailer was also supplied to the British and Canadian Armies, but was used mainly for transporting heavy engineering plant as its maximum carrying capacity was insufficient to cope with the majority of tracked AFVs then in current service.

An interesting special purpose conversion was carried out prior to D-Day. It was realised that, due to the Allied air forces attacks on the railway systems in Europe, a vast amount of reconstruction work was needed to enable the Allies to utilise the rail network for their own supply services, and that material to carry out this task would have to be shipped across the Channel from Britain. A quantity of Federal 604s were therefore converted to ballast body tractors by the removal of the fifth-wheel coupling and the fitting of a rudimentary body 7 ft long by 3 ft $10\frac{1}{2}$ in. wide by 2 ft 4 in. high. This was capable of containing $8\frac{1}{4}$ tons of ballast. The complementary trailer was the British Mk 1 40 ton 24 wheeled transporter trailer modified to carry railway lines. Two wooden baulks were fitted transversely across the trailer, which had had a full floor fitted. Sockets were provided in the baulks into which steel stanchions could be slotted either at full trailer width or 4 ft 6 in. apart. This enabled 30 tons of 36 ft lengths of rail to be transported, the full floor also allowed general cargo to be carried. The Mk 1 trailer was chosen for this conversion as the fixed inner track guides limited its usefulness as a tank transporter.

The Americans found that, due to the purchase of large quantities of Diamond 'T's by the British, an alternative vehicle had to be provided for their own tank transporter units. The Freuhauf Trailer Company were awarded a design contract in 1942 which culminated in the familiar 'Dragon Wagon' or 'M25 Truck-trailer 40 ton Tank Transporter'. The tractor unit designated 'Truck-Tractor M26' produced by the Pacific Car & Foundry Company was unique in its class, as it had a lightly armoured cab accommodating seven crew. This was found to be of little tactical use due to the weight penalty involved, and early in 1944 the M26-A1 unarmoured version replaced it in production. These tractors were interesting, in that they were fitted with the Knuckey chain-drive rear bogie – a single-drive axle drove both pairs of rear wheels via an open chain-and-sprocket system. Twelve forward gears were available via a transfer box, and a Hall Scott Model 440 6 cylinder engine was fitted giving the vehicle a top speed of 28 mph. The engine produced 270 bhp and had a capacity of 1,090 cu. in. The vehicle could carry loads up to 80,000 lb, and the tractor had a forward winch with a capacity of 35,000 lb, while the semi-trailer was fitted with two 60,000 lb winches.

The increase in AFV weights meant that the M15 semi-trailer was becoming overloaded and in 1944 modifications were made to enable it to accept a maximum load of 45 tons, it was then redesignated M15A1. Hinged ramps were fitted to bridge the bogie wheels so that AFVs with a wider track-tread could be loaded.

The M25 remained in service after the War and was still being listed by the US Army Ordnance as a standard vehicle in 1962. By this time, the semi-trailer had been uprated yet again having undergone extensive stiffening of the frame, widening of the platform, installation of inner track guides and removal of the stowage bins. The empty weight had risen to 42,600 lb (M15A1 42,370 lb) but the load on the fifth-wheel had dropped to 13,100 lb (M15A1 14,170 lb), and the bogie now took 29,500 lb (M15A1 28,200 lb). The tyres remained at 14.00 × 24, unchanged since the M15, and were a limiting factor on further development. The trailer was now redesignated M15A2 and the M123 10 ton 6 × 6 truck tractor produced by Mack became the standard prime mover.

The reader will, no doubt, have gathered that the requirements for tank recovery/transportation had been fulfilled by similar vehicles, although in some cases, such as the Diamond 'T' and Rogers trailer combination, the relatively poor cross-country performance rendered them virtually useless for recovery work in the field.

The main factors that govern cross-country performance are:

*1* Adhesion, which can be expressed as the percentage relationship between the driving wheel weight and the gross axle-weight of both tractor and trailer.

*2* Shear, which is the tendency of the ground surface to fail in a horizontal plane under the rotating face of a driven wheel, in simpler terms 'wheel spin'. The shear value of a vehicle can be expressed as a ratio of driving-wheel ground contact area to gross-weight. The higher the ratio, the less chance of wheel spin.

*3* Tyre pressures, these give an indication of ground pressure or flotation, this being the ability to resist settling in soft ground. Single large-diameter 'balloon' tyres being more effective than smaller twin-tyres.

*4* Tractive effort, this is expressed in lb per ton of gross weight and is the maximum power delivered at the driving wheels in the lowest gear-ratio.

Tractive effort, or rather the lack of it, rarely results in poor cross-country performance, and lack of adhesion or onset of shear are more likely causes.

The four main types of tank recovery/transporters already described, are listed below for comparison with the American T8 40 ton experimental transporter.

An unusual design project emanated from Britain at about this time, and was similar in concept to the container handling equipment of today. The idea was that the trailer would be towed into place over the disabled tank, which would then be hoisted clear of the ground by overhead lifting gear and removed for repair in the rear areas. Only scale models appear to have been made of this unique design.

During the same period, the Americans were testing a series of double-ended tank transporters, more accurately described as powered semi-trailers.

The first was produced by Cook Brothers and was, in effect, a tank-carrying platform suspended between two 4 × 4 twin engined power units. The subsequent T4 and T8 vehicles were similar in layout, the T4 and le Tourneau powered wheels and the Mack T8 had four 6 cylinder 240 bhp engines. Its performance figures are given in the previous comparison table.

The other combatants did not produce heavy tank recovery/transporters; instead they relied on heavy cargo trucks to move light AFVs, with towed trailers for tanks up to the 20/25 ton class, although the USSR received transporters from the USA as part of the lend-lease programme.

The advantage of the semi-trailer layout will be readily appreciated by comparing the performances of the two types of Diamond 'T' vehicles. In practice, due to its having large single tyres in its driving bogie instead of the smaller dual wheels, the 6 × 6 configuration of the 'Dragon Wagon' gave it a superior performance. This led to the British WVEE obtaining an M26 tractor for trials, with a ballast body and single 21.00 × 24 tyres fitted all round.

Prototypes of an 8 × 8 tractor were produced by Albion Motors between January 1943 and August 1945. Designated Model CX 33, the first prototype had eight-wheel-drive (8 × 8) with front and rear axles steering, and was powered by two Albion 6 cylinder 140 bhp engines, one of which powered the first and fourth axles while the other powered the second and third. The forward cab housed the driver, who was seated at the centrally positioned steering wheel with a crew seat either side. The winch controls were housed in a similar design of cab at the rear of the vehicle. The engines were positioned side-by-side behind the driver's cab and the rest of the platform was occupied by a ballast body. This tractor was intended to tow a gross load of 75 tons. The load to be carried on one of a new design of trailer being developed to cope with the newer AFV's coming to fruition, was also destined never to enter service.

An 80 ton tilt bed trailer, intended to transport the 'Tortoise', was, in the event, towed by Diamond 'T's in tandem.

The second prototype CX 33, had a revised

| | | Adhesion | Shear | Tyre Pressure | Tractive Effort |
|---|---|---|---|---|---|
| Diamond 'T' & trailer | 6 × 4 | 19.2% | 7.6 | 51.34.90 | 706 |
| Diamond 'T' & semi-trailer | 6 × 4 | 34% | 11.2 | 55.47.60 | 944 |
| Scammell & 30 ton semi-trailer | 6 × 4 | 31% | 10.6 | 25.46.60 | 664 |
| M25 40 ton semi-trailer | 6 × 6 | 63% | 16.1 | 70 | 847 |
| T8 40 ton transporter | 8 × 8 | 100% | 52 | 40 | 1290 |

axle/steering arrangement as the drive to the fourth axle was deleted and the steering transferred to the first and second axles.

One of these tractors was used to tow the Bristol Brabazon at the Company's Filton airfield, and the author noted another in the livery of the Road Research Council towing a road-surface test-trailer at Stonebridge near Coventry in the late 1940s.

# MACHINERY OR MOBILE WORKSHOPS

The problem of maintaining equipment in the best state to take to the field of battle has increased over the centuries in direct proportion to the technical development of the weapons.

Even before the establishment of standing armies, steps were taken to try to ensure that personal arms were kept in a serviceable condition. One of the first recorded attempts to legislate in this manner was an English 'Assize of Arms' issued in 1181. The Assize confirmed the personal responsibilities of a soldier for the state of his weapons and, 'forbade him to sell or pawn his arms and enjoined him to bequeath them to his heirs'.

In later Medieval times, Ordnance Officers (a corruption of Ordinance) were appointed to ensure that personal weapons were maintained in a fit state for war. Central storage and maintenance depots grew up in the main cities; the Tower of London being one of the earliest examples of storage under government control.

Siege trains (and later artillery) needed a certain amount of specialised support in the field, and the first rudimentary mobile workships came into being as horsedrawn forge wagons, wheelwrights and blacksmiths plodding in the wake of the marching armies. The development of mobile repair facilities did not keep up with the progress of the then most sophisticated of arms, the artillery – but the skills and inventiveness of the contemporary armourers and artificers were held in high esteem, and such establishments as the Royal Arsenal at Woolwich were renowned for their craftsmen.

The introduction of the mechanically propelled vehicle into the armed forces of the world necessitated the organisation of proper mobile repair facilities. By 1911, the British Army Service Corps had workshop trailers towed by steam driven traction-engines. Many of these had been converted to motor lorry units by the time World War I started in 1914.

A mobile workshop was allotted to each British Division and consisted of two 'house-type' lorries, one carried stores and spares, the other was the actual workshop vehicle. The latter was equipped with a small lathe, drilling machine, forge and battery charging facilities. Power to drive the machinery and provide lighting was supplied by a $3\frac{1}{2}$ kw 110 v dc generator driven from the vehicle gearbox.

After an initial period of fluid warfare, the static nature of the trench system meant that there was very little call for mobile workshop facilities in France. The situation was different in the other theatres of war, and several mobile workshops served in the Middle East. Two specialised units were formed to support operations in the Balkans. These were truck-borne horse-waggon repair shops, and were fully fitted out with wheelwright facilities, including tyre-shrinking gear and forges, to maintain the horsedrawn transport with which these divisions were equipped.

Between the Wars, development of suitable mobile workshop lorries kept pace with the mechanisation of the British Army. Two principal types of workshop bodies were designed to fit on to the standard 3 ton 6 × 4 chassis. They were designated:

*The 'slave battery' vehicle provided battery charging and servicing facilities for the mechanised units' starter and radio batteries. Emergency starting could be carried out using long 'jumper' leads. This equipment could be installed in most 30 cwt or 3 tonners. A Bedford OY installation is illustrated.*

*Mobile workshops were produced in many types to fulfil a variety of roles. The photograph of a CMP Ford 3 ton 4 × 4 shows how much equipment could be installed in a standard GS lorry.*

*A. Machinery, House Type*
As the name implies, this was of box-like rigid construction, with window panels, hinged at the top, that could be opened along the full length of the body. A platform at the forward right-hand side could be extended to provide a bench for battery charging. As the war progressed this type was gradually superseded by the 'flat floor' type and by modified GS vehicles.

*B. Machinery, Flat Floor*
This body type had several designations, e.g. 'No. 4 Mk 3' and was similar in design to a drop-sided load carrier. The centre section of each side hinged down to form either, a flat platform or, by unlocking a second hinged portion, additional work benches at ground level. The weatherproof canvas tilt was supported on a tubular frame, that could be extended over the extra work areas, and two or more vehicles could be linked together to form a larger undercover workshop facility.

Machinery bodies were fitted onto the Albion BY series, Leyland Retriever, Crossley IGL8, Guy FBAX, Karrier CK6 and Thornycroft Tartar, 3 ton 6 × 4 chassis.

It was the intention to replace these 3 ton 6 × 4 types by the 3 ton 4 × 4 because of the latter's improved cross-country performance, however, due to the losses sustained in the evacuation from Dunkirk and due to the increasing need to equip the expanding army, the 6 × 4 continued in this role throughout the War and for some time afterwards.

Two types of 3 ton 4 × 4 chassis were, in fact, fitted with workshop bodies to supplement 6 × 4 production. These were the Albion FT 11 and the Ford WOT-6, and could carry either a 'Machinery, Flat Floor No. 4, Mk 5' body or a 'Machinery, House Type No. 8' body. The bodies were very similar in design to the earlier versions fitted to 6 × 4 chassis, but varied in detail, as required by the differing layout and dimensions of the 4 × 4 chassis.

When the BEF went to France in 1939, there were two types of machinery lorries in the Port Workshops, eight attached to GHQ Workshops, four with the AA Brigade and twelve in the field. These vehicles provided the following workshop facilities :

Electric and oxy-acetylene welding
Battery charging
Dry air charging for artillery recuperators
Woodworking machines
Electrical component test and repair

*Thornycroft produced about 5,000 Nubian TF/AC4 chassis from 1940 to 1945 and the basic design is still evident in current fire tenders. Illustrated is an RAF tender, type 348, for generating electrical power. It proved popular with showmen in the immediate post-war era.*

*The machinery (house type) body could be fitted out to fulfil several mobile workshop roles. The generator supplying power for the machinery is located forward of the rear bogie. The vehicle is an Albion BY3.*

Milling machines
Drilling machines from $5\frac{1}{2}$in. bench to 32in. radial types
Lathes from $3\frac{1}{2}$in. precision to 8in. capacity.

All lorries with power-driven machinery were initially fitted with a $3\frac{1}{2}$Kw 110v dc generator, driven from a PTO on the gearbox, although, on some earlier vehicles the transmission shaft had to be disconnected and the generator coupled up to the gearbox tail-shaft for static running. This system of power generation had to be abandoned due to, (*a*) failure of the prime mover engine on the move, caused by the oiling up of plugs following continuous stationary running under light load and, (*b*) production of vehicles with PTO could not keep up with the demands of the workshop lorry programme. A normal GS chassis had to be substituted and generator trucks or trailers procured to provide the necessary power for mobile workshops. This problem affected other specialist vehicles requiring electrical power.

Apart from the machinery lorries, there was a wide range of trailer-borne equipment in use. The majority of mobile equipment was carried in 6 ton 4 wheel trailers with drop-sides providing extra working platforms, but the heavier machinery used in the static base workshops was transported on 10 ton low-loading trailers of various types, the base workshops being classed as transportable rather than mobile.

Lightweight workshop trailers were developed for use with the Airborne Divisions, each had to be capable of being towed by a Jeep and carried in a glider, and total weight was not to exceed 1 ton. A complete airborne workshop could be carried on five trailers comprising, (*a*) a 5kw generator trailer, (*b*) a machinery trailer with lathe, drilling machine and grinders, (*c*) a type Z trailer for the repair of radio and associated equipment, (*d*) an electrical repair trailer for MT equipment, this included engine valve regrinding equipment and finally, (*e*) a welding trailer with both gas and electric welding gear. There were two other types also available for airborne use, in an air compressor trailer to provide air at 110 psi for pneumatic tools and a 28 in. diameter circular saw.

The lightweight range of workshop trailers was also destined for service in the Far East, as experience in jungle warfare had indicated a need for smaller vehicles with ability to move along jungle tracks to combat the lack of proper roads.

Also under development for the Far East, was a series of workshop lorries to cope with the moistureproofing of radios/radar, telecommunication and other electrical equipment. These included air-conditioned versions of the radio and instrument repair vehicles, and special drying and spraying workshops where water vapour could be extracted before the electrical equipment was sprayed with water-resistant varnish.

Another prototype was the Turtle, this was a workshop version of the British tracked amphibian 'Neptune' and was intended to form part of Neptune equipped formations with a capability of carrying out front line repairs.

The planned production of workshop vehicles in Britain for 1945 was 3,605 machinery lorries and heavy trailers plus 3,736 lightweight trailers. Production at the time was running at 343 lorries and 190 trailers per month; contracts were terminated with the cessation of hostilities, but not before a total of nearly 450 complete mobile workshop units had been formed for service.

An acquaintance of mine nursed a Ford WOT-6 machinery truck through the NW Europe campaign from D-Day onwards without so much as a scratch, and returned with it to the UK for the disbandment of his unit. He was detailed to deliver his pride and joy to an Ordnance Depot, presumably for storage, and he spent four days checking inventories of tools, making good deficiencies and general-servicing the vehicle. The Ford was delivered on the appointed day, handed over and placed neatly in a line of similar vehicles. The proud owner took a final backward glance, and was mortified to see a huge bulldozer rumble over, and flatten the lot!

The Canadians followed British practice and produced a comparable range of workshop/machinery vehicles on standard CMP chassis. Prior to the production of this range, a workshop body was produced for fitment to the 160 in. wb Chevrolet 4 × 2 3 ton chassis, one of the modified conventional vehicles. This body was designated '5D5', it was similar to the 12 ft steel GS body but had a reinforced floor to cater for the equipment installed. These vehicles saw wide service in the Western Desert.

The early CMP range were carried on a Ford 6 × 4 3 ton Type 60640-N-P chassis, with either the No. 11 or 12 cab. This chassis layout differed from the British 3 ton 6 × 4, as the third axle was not driven, in other words it was a basic 4 × 4 with a third trailing axle. The body was a 14 ft Lindsay house-type constructed of steel panelling, with double doors at the rear, and opening side sections to provide extra working area and overhead cover. There were no side windows. This body was later mounted on the Chevrolet 6 × 6 3 ton Type 60660-M chassis with No. 13 cab, and one version, the 'Artillery Armament Repair Lorry (USSR)' was supplied in large quantities to Russia.

To supplement CMP production, the Canadian QMG Department ordered a special 201 in. wb version of the Diamond 'T' 4 ton 6 × 6 chassis which was known as the Model 969-B and was fitted with a 15 ft Lindsay house-type body containing MT repair equipment. This vehicle normally towed a trailer carrying the penthouse tarpaulins and poles etc. together with the workshop spare wheel and other equipment.

A series of GS type machinery bodies with drop sides was also produced in parallel to the house-type, and fitted to the same 6 wheel chassis. The 8EI was equipped for the Machinery L role and mounted on the Chevrolet 6 × 6 while the slightly larger 8BI (Machinery RE 25 kw) and 8C1 (Machinery H) were carried on the special Diamond 'T'.

In order to rationalise production the QMG Department decided to concentrate on 3 ton 4 × 4 vehicles and drop the 6 × 4 and 6 × 6 from the schedule. This meant that a new series of machinery bodies had to be designed and produced, these had to fit on to the smaller chassis while fulfilling the identical requirement for machinery facilities. Both the Ford F60L and Chevrolet C60L 158 in. wb 3 ton 4 × 4 chassis were used, and a 12 ft house-type body as well as the strengthened GS type were available; the intention being to drop the house-type from production in order to concentrate on GS production. These bodies were known as the Mk I series in the 5F (GS type) and 5J (Lindsay house-type) classes and were equipped to fulfil a variety of roles.

At the lower end of the range was a signals repair workshop known as the 'Machinery ZL' installed in the standard Chevrolet C8A 4 × 4 heavy utility truck, which had body type 1C10.

Special mention must be made of the mobile welding plant, designated 'Machinery KL' by

*Various types of workshop equipment could be installed in the GMC truck $1\frac{1}{2}$ ton 4 × 4 ordnance maintenance. British modifications included fitting POW can carriers forward of the spare wheel.*

the British and Canadians, it was carried on 15 cwt 4×2 or 4×4 chassis. The welding equipment was of the 300 amp electric type, powered by a separate engine and it made available the usual welding equipment such as a vice, table, welders screen and portable grinder together with a selection of electrodes for both normal steel and armour plate welding.

Unfortunately, the engine installed in the British version proved to be incapable of providing the 300 ampere necessary for continuous welding (400 amp peak) and either the Willys MB (ex-Jeep) or Ford V8 engines were substituted – the Canadians already used the V8 in their version of the KL.

Pre-war planning had completely failed to appreciate the extent to which welding facilities would be required in the field, and equipment was inevitably in short supply. Every use was made of captured equipment but it was realised that nearly 40% more power was needed when using enemy equipment, this was due to the thicker flux deposit on German welding rods.

Apart from general cutting and welding work in workshops, there was a constant call for assistance in straightening out chassis frames and axles and for modification work. The welding of armour plate presented special problems, as variations in local heating and cooling caused internal stresses to be set up, resulting in loss of protection against shot penetration. Methods of repair had to be evolved as tank crews disliked repair patches welded over damage even though these gave adequate protection, and it was more usual to make good holes and gouges by filling them with deposited weld rod and grinding flush. In some cases, if the shot had not penetrated it was welded firmly into position and left *in situ*. Thin armour had to be patched as filling was impracticable, and cracks were normally welded over.

Welding vehicles were attached to recovery units as their services were often needed to cut away damaged tracks etc. before a tank could be moved by the recovery vehicle. The gas welding equipment carried by recovery vehicles was not always adequate.

During the rapid advance to Brussels and

Antwerp by three British armoured divisions following the beach head break-out, it was customary practice to leave broken-down tanks by the roadside, and tack weld all hatches in the closed position to prevent looting etc., until the hard pressed recovery and repair units caught up and retrieved them.

The Americans initially had two types of workshop vehicles in production in 1939, both were designated ' Truck Ordnance Maintenance $1\frac{1}{2}$-3 ton 4 × 4', and both were cab-over-engine (COE) designs with bus bodies, produced by GMC. The first was the Model AFKX-352-8A and this was supplemented by the –8G, being produced to fulfil the following roles:

| | | | |
|---|---|---|---|
| Truck, small arms repair | M1 | * | (SM 2521) |
| Truck, artillery repair | M1 & M2 | | |
| Truck, automotive repair | M1 & M2 | * | (SM 2518) |
| Truck, instrument repair | M1 | | |
| Truck, light machine shop | M3 | | |
| Truck, machine shop | M4 | * | (SM 2520) |
| Truck, spare parts | M1 & M2 | * | (SM 2522) |
| Truck, tank maintenance | M1 | * | (SM 2524) |
| Truck, tool & bench | M2 | * | (SM 2523 & 2545) |
| Truck, welding | M3 | * | (SM 2526) |

** These models were supplied to the British Army, contract reference numbers given in brackets.*

The eventual standardised 'Truck Ordnance Maintenance $2\frac{1}{2}$ ton 6 × 6' was a house-type Model ST5 (non-collapsible) body mounted on the GMC CCKW-353 chassis, originally with a closed cab but, in common with other US military trucks, changed to the familiar open style Model 1619 cab from July 1943. The basic body, like its predecessors was equipped to provide a variety of workshop facilities, these were:

| | |
|---|---|
| Truck, artillery repair | M9 & M9A1 |
| Truck, automotive repair | M8 & M8A1 |
| Truck, electrical repair | M18 & M18A1 |
| Truck, instrument bench | M23 |
| Truck, instrument repair | M10 & M10A1 |
| Truck, machine shop | M16 & M16A1 |
| Truck, small arms repair | M7 & M7A1 |
| Truck, spare parts | M14 |
| Truck, tool & bench | M13 |
| Truck, welding | M12 & M12A1 |

This chassis was also fitted with a cargo-type workshop body and used by the US Corps of

*A South African Army mobile workshop in action.*

Engineers as the 'Truck, Shop, Motorized, $2\frac{1}{2}$ ton 6 × 6'. It was capable of providing facilities for general and electrical repair, light machine shop, small tool repair, tool & bench and welding.

The house-type body continued in service for several years after the end of the war, but was 'non-classified' although the number of roles had increased and by then it was known as the 'Truck, Shop, Van $2\frac{1}{2}$ ton 6 × 6'.

An improved version, the M 535, had a so-called 'collapsible body', in actual fact the roof section could be lowered to reduce the overall height by 2 feet.

A number of Studebaker US6–U3 162 in. wb $2\frac{1}{2}$ ton 6 × 6 trucks were supplied to the British Army under lend-lease on Contract SM 2440 with 12 ft bodies equipped as a general workshop.

The Germans mounted the majority of their workshop bodies on either, the medium 4 × 2 Mercedes-Benz L3000S/066 or, the 4 × 4 Opel 'Blitz' 3.6–6700A, choosing like the Allies their most readily available vehicle. They used a standard house-type or van body. This could be equipped to fulfil a wide variety of roles ranging from the Kfz 305/1 teleprint truck (FernschreibanschlussKraftwagen G), the Kfz 305/83 motor transport repair truck (Kraftwagen workstatt-Kraftwagen), the Kfz 305/111 repair shop truck for locksmiths and welders (Werkstatt-Kraftwagen für Tischler, Maler, Sattler) to the Kfz 305/137 Luftwaffe repair truck.

Apart from this standardised range, there were numbers of other workshop vehicles in service, such as the Kfz 79, a house-type mounted on the medium 6 × 4 chassis. The Kfz 42 signals workshop was a house-type fitted to the medium 4 × 2 chassis range, as was an armourer's workshop, except that this was a converted cargo-type body. Finally, the Kfz 51 house-type general workshop was fitted to pre-war heavy 4 × 2 commercial chassis.

# AMPHIBIANS

Many countries had turned their hand to designing and testing amphibious tanks in the 1920s and 1930s but these were, for technical reasons, based on existing light tanks and could only be used for short river crossings in calm water. There were no supporting amphibious vehicles, and reliance was still placed on light assault boats and bridges to overcome water obstacles.

Some experimental work was carried out in Germany in the 1930s following the appearance of a 10 × 10 amphibious prototype vehicle by Büssing-Nag in 1929.

An amphibious tracked tractor was produced in 1934. It was designated 'Land-Wasser-Schlepper' (LWS) and was intended for use as a general purpose tractor on land, doubling as a tug for hauling and manoeuvring rafts and pontoon bridging equipment.

A refined version saw service in World War 2 and had accommodation for 20 men in the cabin but there was no room for general cargo. Performance on land was limited as the tracks were only really intended to assist in entering or leaving rivers and for crossing shoals or shallow water. When afloat the LWS was a graceful craft, but it was most ungainly on land and could not be classed as a true amphibian.

An LWS captured by the British was tested by the WVEE, and when run for a few seconds at full throttle in the test tank its propellers succeeded in pushing hundreds of gallons of water up the ramp and out of the tank. Unfortunately, it was then found that the vehicle had insufficient track grip to negotiate the ramp out of the tank. Incidentally, the 'funnel' was a combined up and down draught ventilator.

The Trippel SG6/38 was a true amphibian, it first appeared in 1935, entering full military service in 1938. This version had a box-like rear body with a bulbous forward engine compartment and was powered by an Opel 6 cylinder $2\frac{1}{2}$ litre engine. The engine had to be modified to cope with steep angles of entry and departure from water by extending the sump and rearranging the oil pump to ensure that the lubrication system functioned satisfactorily at these extreme angles. A Sirocco-type fan provided cooling air to the engine, and also partly pressurised the compartment helping to keep it watertight. The propeller was driven from the main gearbox via its individual gearbox with a forward and reverse gear, then by enclosed triplex chain to the short propeller shaft rather similar to an 'inboard-outboard' boat drive. For land travel the propeller could be swung sideways to give adequate ground clearance.

This vehicle could carry 1 ton overland and 2 tons when waterborne. A more refined version, the SG6/41, was introduced in 1941 with a more boat-like hull and continued in production until 1944, with subcontract by Bugatti.

The other German amphibian of the period was the well known Schwimmwagen originally designed and produced by Porsche, the Volkswagen VW 166 Kfz 1/20 became the standard production version and over 14,000 were produced between 1942 and 1944.

The rear-engined layout was an added bonus for the amphibian as the weight at the rear kept the nose up when entering the water, and an excellent angle of approach (to 60°) was thus possible. A driven front axle was added – a necessity for amphibians, the only other alternative being to leave the water stern first – and a special emergency low gear was added to the front of the normal gear box, which connected to, and was the only gear available to, the front axle. The propeller could be swung upwards out of the water, and was driven via a chain assembly from a dog at the end of an extension to the engine crankshaft, thus there was only forward motion available at engine speed and no neutral. Steering was by the front wheels and a maximum water speed of 6 mph was possible.

These German vehicles were essentially for crossing rivers in relatively calm conditions and could not, in any way, have been considered as suitable for seaborne launching from landing craft.

The US Marine Corps was probably the only marine force in the world to actively pursue the goal of true amphibious capability between the wars. This policy was to pay rich dividends in the island hopping Pacific campaign after the trauma of Pearl Harbor.

One of the first amphibious vehicles to draw the attention of the Marine Corps was the Roebling tractor. Donald Roebling started on the design and construction of the 'Alligator' in 1932, with the intention of providing a vehicle capable of negotiating the swampland of the Florida Everglades for search and rescue work. Steady development continued during the remainder of the 1930s; and by 1937, a water speed of 9 mph had been achieved by track propulsion which was not exceeded by the later LVT series.

Interest in the Alligator by the Marine Corps led to the acquisition of three test vehicles in 1940, and later in the same year, a production order was placed for 200 vehicles. The main change made in production was the use of steel instead of aluminium alloy, the vehicle was fitted with 150 bhp Hercules engines. Initial deliveries commenced in July 1941, the vehicle being designated LVT-1. As there were problems with track and suspension which resulted in thrown tracks, re-design was undertaken by the Borg Warner Corporation and the LVT-2 (the Water Buffalo) superseded the LVT-1 on the production line in 1943.

The LVT-2 incorporated many components of the M3 light tank including the Continental 250 hp 7 cylinder air-cooled radial aero-engine and improved track grousers. One of the main problems with the LVT was the rapid wear of grousers on land, and their replacement was a continual maintenance penalty. A drawback with the early LVTs was that the location of the engine in the rear prevented the installation of a loading ramp.

The Borg Warner Corporation carried out a further re-design in which the engines were moved into the side sponsons, and a rear ramp added, enabling vehicles like the Jeep to be carried. The new version, the LVT-3, utilised M5 light tank components and remained in service for many years after its introduction in 1943.

Another re-design of the basic LVT was carried out by the Food Machinery Corporation, the engine being positioned forward, leaving a wider hold and wider rear loading ramp. The greater carrying capacity and ease of loading led to the wide adoption of the LVT-4.

The success of the LVT series led to many modifications, as well as conversions to other roles than that of straight cargo carrying. The obvious vulnerability when carrying out opposed landings, resulted in bolt-on armour being fitted to the front and cab. Armed variants were produced that were, in fact, amphibious tanks capable of giving close fire support to the initial assault troops.

The US Marine Corps and Army in the Pacific were given priority, and LVTs took part in island assaults from Guadalcanal in 1942 to Balikpapen in July 1945. The British had watched the success of the LVT in the Pacific campaign with great interest and not a little envy, however, it was not until late 1944 that quantities of LVT-2s and LVT-4s were delivered to the British Army under lend-lease.

Some 200 LVT-4s were delivered to the British Army in Italy in February 1945, and an amphibious RASC regiment was formed from existing units including 385 and 931 GT Companies, 15 Army Transport Column and HQ RASC. It was the intention to use this regiment in the flooded and swampy area around Lake Comacchio, and only six weeks were available for training. The five squadrons of the regiment were capable of transporting a complete infantry brigade consisting of the brigade HQ, three infantry battalions, two troops of 25 pounders, an engineer squadron and a recovery section. As there were no special purpose LVTs included in the delivery, local modifications were designed and carried out by the REME 686 Infantry Workshops at Naples and Acona. These modifications included fitting seats for infantry in 140 'fantails', as the LVT-4 was codenamed in Italy. 144 sets of stretcher equipment were manufactured and 156 radio control boxes had to be repositioned, and 10 vehicles had a second radio set fitted, as did 15 command vehicles and 8 vehicles intended for navigational leaders.

Other major modifications were carried out, these were:

*1* 25 pounder portee. Bailey Bridge transoms were used to form supports for the No. 9 firing platform and the trail, together with additional intercommunication equipment and special loading ramps.

*2* Sturgess ramp. Two LVTs had special ramp carriers fitted together with a winch which enabled two men to complete a launching in about five minutes.

*3* Recovery vehicle. A cut down Chevrolet gun tractor was mounted on two Bailey Bridge cribs secured to amphibian floor. A jib was mounted on the chassis with a 3-to-1 block and tackle. The rear brake drums of the gun tractor were used as bollards for securing or warping the LVT.

*4* Mat layer. This was similar to that devised by 79th Armoured Division. The mat was 15 ft wide and 60 ft long and was constructed from coir matting reinforced by 2 in. steel tubes every foot. The end of the mat was fed under the tracks and uncoiled as the vehicle moved forward.

*5* 2 pdr and Jeep transporter. The 2 pdr was secured in a similar manner to the 25 pdr, and the Jeep was carried above it on a wooden structure reached by special wooden loading ramps. Operations supported by the amphibious regiment started successfully with the capture of the Menate Gap by the Queen's Royal Regiment.

Both the LVT-2 and LVT-4 were operational with 21st Army Group in NW Europe, equipping regiments of the 79th Armoured Division in time for the clearance of the Scheldt Estuary which started in October 1944. These regiments were in constant demand to support operations in the flooded areas of Holland, culminating in the Rhine crossing. As in Italy, special field modifications were carried out to improve the efficiency of the LVTs which were known in this theatre of war as 'Buffaloes'. There were two main special purpose vehicles:

*1* 17 pdr carrier. LVT-2s were fitted with special ramps for loading and carrying the 17 pdr anti-tank gun; this was rather top heavy, but was nevertheless successful in getting heavy anti-tank weapons forward with the assault waves.

*2* Carpet layer. This was similar in principle to the mat layer used in the Italian campaign. It was mounted on the LVT-4, and was developed because operations in the Scheldt revealed that DD Shermans had difficulty in climbing steep muddy river banks.

Other modifications included the fitting of appliqué armour, extra armament and gun shields. Some 400 LVTs were fitted with a 20 mm Polsten cannon as bow armament in preparation for the Rhine crossing, and this work was carried out in eight weeks. Two innovations were also used during this operation, the first was codenamed 'Tabby' and was an infra-red low visibility driving aid, using infra-red head lights, the driver viewing the projected images through special binoculars; this principle was already in use by the Germans as a sighting system for small arms. The system enabled LVT drivers to drive through smoke-screens and battle fog as well as at night, and was intended for wide scale use in the Far East if hostilities had continued.

The other device was a navigational or direction-keeping aid consisting of two No. 19 radio sets, positioned at a measured distance apart on the launching bank of a river crossing and transmitting signals which were picked up by the LVT radio. Direction could be maintained to within $\frac{1}{2}°$ despite the fact that equipment was still in the experimental stage at the time of the Rhine crossing and Operation Plunder.

LVTs remained in service with many armies long after the end of World War 2, and the LVT-3, the principal amphibious vehicle of the US Marine Corps in both armed and standard form, exerted strong influence on the British designed Neptune.

The other tracked amphibian to enter production in the USA, was the 'Carrier, Cargo M 29 Weasel'. This was a diminutive vehicle compared to the LVT, but was designed for an entirely different role, that of supporting airborne troops in snow covered terrain. It was developed from the T15 and T24 prototype. The M29 had a very low track pressure of 2.1 lb per sq in. at zero penetration when fitted with 15 in. tracks, and had a very good performance over snow, mud and swamp. The vehicle was capable of operating in temperatures from −40°F to +120°F (−40°C to +49°C) with a crew of four, or a crew of two plus an 860 lb payload, in either case, with a towed load ranging from, 1,200 lb under the worst conditions to, 2,400 lb in the most favourable conditions. The M29 was also airportable in the C47 (but not with later 20 in. tracks) and could negotiate smooth water at slow speed. Its amphibious capabilities, however, were limited by its small freeboard and lack of water steering. Propulsion in the water was by the tracks and was not very affective.

In order to improve the 'amphibious' capabilities of the M29, the Studebaker Company added bow and stern watertight cells, sponson air tanks, track side-panels, cable controlled rudders, power capstan, hand bilge pumps and a surf-plate. Its 20 in. tracks gave the M29C (as this version was designated) a ground pressure of 1.93 lb per sq in. at zero penetration, despite an increase in all-up weight from 4,925 to 6,040 lb. One disadvantage was that, due to the hull extensions, the angle of approach was reduced from 90° to 47°, and the angle of departure was reduced from 60° to 36°. The increased buoyancy of the M29C gave a freeboard of $10\frac{1}{2}$ in. at the bow and 8 in. at the stern at maximum weight.

Water propulsion was improved, by the upper or return run of the now enclosed track forcing water forward against the bow cell, then downwards and outwards through an orifice, thus giving a form of water jet thrust. The Weasel could still operate only in calm slow moving water, but the speed was increased to 4 mph.

Deliveries to US operational units began in mid 1944, and the British Army received lend-lease vehicles later in the year, in time for

operations in the Scheldt Estuary, Lake Comacchio and River Po areas.

These vehicles did not form the sole equipment of units, but were issued as part of the establishment of amphibious regiments equipped in the main with LVTs. The Weasels were utilised for command, radio, ambulance, signal line laying and light cargo duties, in fact, they were the maid of all the work undertaken by the Jeep in normal units.

For the assault on Walcheren, a complete platoon of Weasels formed part of 529 GT Company of the RASC in support of the British 52nd Infantry Division, and others served with the amphibious regiments of the 79th Armoured Division.

Weasels were also destined to play a useful part in the Far East and Pacific campaigns where their performance over swamp was put to good use. It was not until after the War that the Weasel was used in the role for which it was intended, that of an oversnow vehicle, in service with the Norwegian and British Armies protecting the Northern flank of NATO.

The Americans also produced wheeled amphibians that could be loosely compared to the LVT and M29, these were the DUKW and

*The GMC Dukw-353 universally known as the 'Duck' proved its worth in every theatre of the war. This example is in 'showroom condition' bearing a post-war census no. 71YP37 and 79th Armoured Division sign.*

amphibious Jeep.

The combination of GMC model-code letters that resulted in DUKW was a fortunate coincidence for an amphibious truck; although unkind comparisons were made when it was likened to a real duck – in that it appeared serene, but had to paddle like hell below the surface to make any headway.

The combination of the GMC 6×6 chassis and the Sparkman & Stephens body made a very efficient vehicle despite the lack of a rear ramp, and many thousands performed yeoman service during and after the War in many countries.

The DUKW utilised the automotive parts of the GMC $2\frac{1}{2}$ ton CCKW truck, with drive to the propeller from the three lower gears giving a water speed of 6.4 mph.

The available cargo space was 196 cu ft and the maximum load was 5,000 lb. A self mounted jib was available for loading, this was originally intended for the US 105 mm howitzer. There were several experimental and special versions of the DUKW, ranging, from the spectacular cliff scaling model with Merryweather fire-engine turntable ladder, to the 25 pdr portee – this was, as far as can be ascertained, never actually fired, which was probably a very wise decision.

The first DUKW operations were in New Caledonia in March 1943, followed by the North Africa landings shortly afterwards. In preparation for the invasion of Sicily, the first British DUKW companies were formed as part of the RASC. In fact, the organisation was identical to a normal truck transport company and consisted of 120 vehicles plus 12 in reserve. This meant that units could be converted with a minimum of disruption: merely a change of equipment, one for one, from amphibious to land role. Some 230 DUKWs plus a few amphibious Jeeps were allocated to the British 8th Army and the majority were only delivered to No. 385 GT Company RASC six weeks before the invasion date. A total of approximately 1,000 DUKWs took part in the invasion of Sicily, supporting the American, British and Canadian forces, and carried out their intended role extremely efficiently. However, the first signs of misuse by commanders became apparent when DUKWs were sent inland to deliver supplies instead of unloading on the beaches. This error was committed time after time and it resulted in traffic jams as the ungainly DUKWs were routed along roads where their lack of manoeuvrability and size caused unnecessary obstruction.

DUKWs were successfully used across the Straits of Messina in the invasion of Italy, despite the turbulent waters of this notorious passage, with the first DUKWs unloading on the beach 90 minutes after H-hour.

After the initial landing, the DUKWs carried on re-supply along the coast as the advance continued, then took part in the landings at Salerno and Anzio and assisted in delivering supplies to Naples until the docks could be made operational. Later, amphibious companies played a major part in re-supplying bridgeheads across the Sangro, Volturno and Garagliano rivers. Some of these journeys involving a sea voyage of up to 12 miles.

The next major operation was the invasion of Europe and up to 2,000 DUKWs were in use by the American and British. Once again, a lack of conventional trucks on the beaches led to DUKWs travelling further inland than was planned. The DUKWs were ferried across the English Channel in LSTs, or suspended from ships' davits. The weather was far from favourable, but losses from weather and enemy action were lighter than anticipated. In the British Sector, DUKWs were handling over 10,000 tons of stores a day, and *in toto* passed more tonnage than came ashore via the Mulberry harbours. It is reported that one DUKW accidentally exploded a mine dropped by night, and was awarded the traditional bottle of Scotch – the Royal Navy's award for mine sweeping! The eleven DUKW companies of the British 21st Army Group were reduced to two companies after the break-out from the bridgehead. The remainder reverting to conventional transport companies.

The next major DUKW operation was in the flooded areas of the Scheldt Estuary and the Reichswald. Roads were flooded to a depth of 3 ft. However, by drafting in a second complement of drivers, the DUKWs were utilised round the clock. Enemy action and operational conditions took their toll, and tyre damage caused by unseen obstacles became a serious problem. Support was provided by 10 ton lorries waterproofed to enable them to operate in the flood water. Preparations were then put in hand for the Rhine crossing, and 200 DUKWs were refurbished by the British for this operation. The Americans making similar preparations. The crossing was not as difficult as that over the lower Maas, due to the slower current; but some of the less experienced drivers had difficulty in negotiating the steep muddy bank exits. In fact, although coral reefs played havoc with tyres, and soft volcanic-ash beaches found on some islands, such as Iwo Jima, proved difficult to negotiate, mud was the worst enemy of the DUKW.

The Americans made good use of DUKW companies in the Pacific from their first baptism in July 1943 in New Guinea until VJ Day. Up to 13 companies being in action at Leyte, and Australian Army DUKW units fought alongside their American allies. The good seagoing characteristics of the DUKW, particularly in heavy surf, earned them a high reputation in this theatre of operations.

DUKWs were also in limited use by South East Asia Command (SEAC), and an amphibian training school was established at Cocanada in India. One of the DUKW companies to see extensive service was No. 387 Company RASC, organised into two platoons with 20 amphibians each. Because of the distance this company often operated from base, and because of the ever present dangers of damage from floating debris as well as enemy action, one DUKW was fitted out as a workshop vehicle. Rivers in Burma were often swift flowing and conditions marginal for safe DUKW operation, with currents of 4 to 5 knots. In these conditions a downstream journey of just over 30 miles took, on average, 4 hours while the return trip could take anything from 16 to 20 hours. Great use was made of DUKWs in moving troops and vehicles across or down the Rivers Chindwin and Irrawaddy, by towing them on rafts. DUKWs also assisted bridging operations by towing sections of Bailey bridges into position, they did this for the 1,000 foot bridge at Kalewa. The Irrawaddy was a formidable water obstacle about four times as wide as the Rhine, littered with sand bars and $2\frac{1}{2}$-knot currents, but this did not prevent a depleted company from moving over 6,000 men, 200 vehicles including AFVs, guns and supplies across the river in 72 hours. There can be no doubt that without the DUKW, operations in Burma would have been very different indeed.

Many DUKWs were also supplied to the USSR under lend-lease and were highly regarded by the Red Army. The tactical merits of the amphibians were probably highlighted by Russian experience with the DUKW, which led them to develop their own version, the ZIL-485, incorporating an enlarged cargo hold and having the advantage of a loading ramp at the rear. These were followed in post-war years by a succession of amphibians of all types.

The Americans also produced an amphibious version of the Jeep, probably influenced by the German Schwimmwagen.

The firm of boat designers Sparkman & Stephens, of DUKW fame, designed a boat-shaped hull to fit the Ford GPW chassis and this model was known officially as 'Truck, Amphibian, $\frac{1}{4}$ ton 4 × 4 Ford GP-A', unofficially it was known as the Seep. Ford were awarded contracts for 12,778 vehicles, production commenced in September 1942, but was terminated in October 1943 before the full quantity had been produced as the Seep did not live up to American expectations.

Due to the extra weight of the amphibious body, an additional leaf was fitted to each road spring. The gross weight was 4,460 lb with a payload of 800 lb. Water propulsion was by a propeller driven from a PTO on the transmission. A maximum water speed of 5.5 mph in 2nd gear was permitted, but the range was then only $18\frac{3}{4}$ miles as opposed to 250 miles on land. A rudder provided water steering, a small anchor, bilge pump and powered capstan were also fitted.

Apart from service with the US Army, amphibious Jeeps saw service with the British and Russian armies, usually as part of the complement of DUKW companies for reconnaissance and liaison duties. The Russians developed their own version, the GAZ-45, which remained in service for a number of years after the war.

The British entered the field of amphibian design and production rather later than the other combatants, and then mainly as an insurance policy in case supplies of suitable lend-lease vehicles failed to meet requirements.

The first tracked amphibian to be built was the Morris-Commercial Argosy and the prototype appeared in 1942. The Argosy superficially resembled the German LWS, a captured example having been tested at the WVEE, but the Argosy was a cargo carrier rather than a tractor. It was powered by a Nuffield Liberty 340 bhp V-12 similar to the unit fitted in cruiser tanks. The track and suspension bore a close similarity to Valentine components, and its seagoing characteristics were very good. However, despite its 9 ton payload, the Argosy did not enter production because the lack of a loading ramp and the vehicle's weight restricted its operational usefulness.

A second and more promising design came from the same stable and was obviously influenced by the US LVT series. At the end of May 1944, the British War Office awarded a contract to Morris-Commercial for the design of a tracked amphibian, to be loaded by a rear ramp, with a hold big enough to contain a 17 pdr anti-tank gun or a 3 ton 4 × 4 truck. Thus, although similar in layout to the LVT-4, the Neptune, as the new amphibian was named, was a much larger vehicle. Other features were its ability to cope with 12 ft surf, and carry a payload of 5 tons with a water speed of 5kn, and a maximum spearheadland speed of 18 mph. It could be operated by a crew of two and all controls, including those for winches and ramp, could be operated from the driver's seat. The engine was a Meadows 12 cylinder developing 280 bhp. The prototype Neptune was completed and ready for trials before Christmas 1944, a remarkable feat even by wartime standards. Although production started, and a war establishment for a Neptune equipped amphibious regiment was drawn up, including a flame-thrower version known as the Sea Serpent and a workshop christened Turtle, contracts were cancelled at the end of the war.

The policy of complementing American vehicles by British designs produced an equivalent to the DUKW or nearly! The Terrapin Mk 1 Amphibian 8 × 8 4 ton was designed by Thornycroft to a specification formulated by the TT2 Department of the Ministry of Supply. It was intended as a ship-to-shore transporter of netted cargo up to 4 long tons, as opposed to the DUKWs $2\frac{1}{2}$ short tons. The driving position was amidships together with engines and intake trunking, the driver was protected from slung loads by two tubular rails. The forward hold had a capacity of 112 cu ft and was protected by a folding canvas spray screen, the rear hold took slightly less. The water speed of the Terrapin was 5 mph fully laden, and 15 mph on land. The 8 × 8 layout was somewhat unique, in that, each of the two Ford V-8 85 bhp engines drove the wheels on the side that the engine was mounted, and steering on land was by juggling the engine throttles. Only the middle two pairs of wheels were normally in contact with the ground, otherwise the method of steering would have caused excessive tyre scrub. The raised front and rear wheels gave improved angles of departure and approach. The overall length of the Terrapin was limited to 23 ft, this was so it could fit into the lifts of landing ships (Landing Ships, Tank [LST]). The usual marine accessories were fitted, including constant running bilge pumps, anchors and towing bridles. The Terrapin was intended to leave a landing craft ramp stern first to clear the propellers and rudder, and ramp approach was made conventionally. Production of the Terrapin was transferred to the Morris-Commercial works at Adderley Park, where delivery started in late 1943 of over 500 vehicles. DUKW drivers were re-trained on the Terrapin, and the 1st Assault Brigade RE of the 79th Armoured Division received the new amphibian in time to take part in the operations to clear the Scheldt Estuary during the latter quarter of 1944. The lack of a rear loading ramp restricted the Terrapins operational usefulness, and an improved version, the Mark II, was designed by Thornycroft. The wheel layout, transmission and engine was similar to the Mark I, although the cab and engine compartment had been moved forward to leave space for a large unrestricted hold to the rear of the vehicle. The Mark II was 5 tons heavier and 7 ft longer than the Mark I, but only prototypes were produced.

Another Morris amphibian prototype was the Gosling, designed by Alec Issiganis of 'Mini' fame. This was, in fact, a motorised handcart with a single Villiers motorcycle-type engine which drove the rear wheels or an outboard propeller.

About the time that the Terrapin Mk 1 was undergoing trials, a strange looking prototype appeared as a private venture. This was the Oppermann Scorpion which was an 8 × 8 of unusual wheel layout. There was a wheel fore and aft on the centre line, while a larger wheel either side mid-way along the hull overlapped smaller diameter wheels in front and behind. This vehicle did not enter production.

Amphibians had to meet strenuous operational requirements and, because of the nature of their employment immersed in salt water, needed constant maintenance. Tracked amphibians like the Buffalo had a very short track life when used on hard surfaces.

# GUN TRACTORS

This class of vehicle can, for convenience, be divided into three classes, Field, Medium and Heavy Artillery. Generally speaking, the Americans did not produce special gun tractors except for the heavier pieces. The Germans used semi-track tractors, and the British had a range of special purpose wheeled tractors.

The problem with providing a suitable tractor lay in the differing requirements of various artillery pieces, the size and type of ammunition, number in crew, laying equipment, etc.

The Americans solved this by using standard cargo trucks with 'bolt on' attachments. The $1\frac{1}{2}$ and $2\frac{1}{2}$ ton 6 × 6 was utilised as a tractor for the 105 mm and 155 mm howitzers. Seats for the gun crew were already available and there was ample room for personal and special kit. Ammunition was stacked in the body, secured by chocked rails. As no limber was provided, ammunition had to be dumped at the gun position and the conspicuous tractor moved off the gun line, this was an inconvenience rather than a great disadvantage.

The Mack NM6 6 ton 6 × 6 and NO2 $7\frac{1}{2}$ ton 6 × 6 were used as prime movers for heavy artillery such as the 155 mm gun and 8 in. howitzer. The NO2 was the only one to have special equipment fitted as standard, a large tubular bow, carrying a chain-hoist, was mounted at the rear to lift the standard US heavy artillery trail-clamp into position. When used in this role, a towing hook was not fitted.

Several other special artillery prime movers were built in small quantities, but they never achieved widespread service due to operating and maintenance costs and lack of stowage space.

*Crossley produced a field artillery tractor with Kegresse bogie based on their light 6 wheeler (30 cwt) chassis. The winch rollers and combination of acetylene and electric headlights are notable features.*

The M3 half-track was often used as a tractor for anti-tank guns and field artillery but, in mid 1943, a series of fully tracked tractors were introduced which were intended to replace all previous types. These tractors were, 'Tractor, High Speed, 18 ton M4' for towing the 3 in. and 90 mm AA guns, 155 mm gun, 8 in. and 240 mm howitzers, 'Tractor, High Speed, 13 ton M5' for handling the 105 mm, 155 mm and 4.5 in. pieces, and, 'Tractor, High Speed, 38 ton M6' which replaced the M4 as a tractor for heavy artillery. All these vehicles had a multiple stowage arrangement so that conversions to cater for various sizes of ammunition could be easily and speedily carried out. Both the M4 and M6 were supplied in small quantities to the British Army, together with 155 mm guns and 8 in. and 240 mm howitzers.

Converted tanks and tank destroyers were often used as gun tractors but were not as practical as purpose-designed tractors.

The M6 was a large and powerful vehicle as befitted its task. It was powered by two Waukesha Model 145 GZ 6 cylinder petrol engines of 815 cu in. capacity, and each developed 215 bhp at 2,100 rpm. Both engines were mounted together, forming a power pack with the twin clutch assembly. Each engine drove a fluid torque converter via the clutch, and the torque converters were coupled into a single transmission and steering unit. Torque converters were fitted, as this was a more efficient way of transmitting optimum power when starting from rest with a heavy load in tow. Maximum allowable speed was 20.5 mph and average fuel consumption was 0.4 miles to the US gallon, a 300 gallon fuel tank giving a range of between 100 and 125 miles with a towed load. Maximum loaded weight was 76,000 lb including a payload of 16,000 lb. Up to ten crew could be carried, and ammunition stowage was provided for the following:

Either 32 round and 32 charges for 4.7 in. M1 AA gun
or 14 „ „ 14 „ „ 8 in. M1 gun
or 14 „ „ 14 „ „ 240 mm M1 howitzer

A pivoting hoist with chain block was provided for handling ammunition.

A Gar Wood 6 m winch was fitted, capable of exerting a pull of 55,000 lb. Both compressed air and Warner electric brake couplings were fitted. Overall dimensions were 21 ft $5\frac{13}{16}$ in. × 10 ft $\frac{1}{2}$ in. × 8 ft $8\frac{1}{16}$ in. to top of gun ring. The M6 was capable of towing its load up a slope of 30 degrees providing it could maintain traction.

The British Army had evaluated many different types of gun tractors between the World Wars and even carried out a brief flirtation with self-propelled artillery in the shape of the Birch gun. However, financial considerations led to the demise of SP artillery and efforts were concentrated on towed guns. Three types of tractors were developed, fully tracked, semi-tracked and wheeled. Small numbers of each were ordered and subjected to prolonged service trials in the late 1920s and early 1930s.

The fully tracked vehicles were known as Dragons, the light Dragon series being used to tow field artillery, mainly the 18 pdr and 3.7 in. howitzer. The medium Dragons were issued in small numbers to Medium Regiments of the Royal Artillery to handle the 60 pdr. Later versions of the light Dragons saw service with the 1939/40.

Trials with the Citroen-Kegresse semi-tracked tractor led to a similar vehicle being produced by the Crossley Motor Company as a tractor, load carrier and battery car, following smaller quantities of Burford semi-tracks into

*This Burford-Kegresse MA 3 ton field artillery tractor was fitted with the later positive drive type of Kegresse track.*

service. The alternative half-track design from the British Roadless Traction Company was used in Morris, FWD and Guy designs; again, these were produced in relatively small quantities.

Improvements in the design of all-wheel-drive vehicles, coupled with the expensive procurement and maintenance costs of the semi-tracked vehicles, sealed the fate of the latter. The successful Morris-Commercial 30 cwt 6 × 4 C and D types were developed into the CD/SW 6 × 4 field artillery tractor of 1935. This vehicle had the classic British 6 × 4 layout using the WD-type patent twin-axle rear bogie and could be fitted with overall chains. Powered by a Morris 6 cylinder 3,485cc side-valve petrol engine developing 60 bhp, the CD/SW was capable of negotiating a slope of 1 in 2½ under the most favourable conditions. A PTO driven 4 ton winch was fitted with 120 ft of 2 in. steel cable, and could be used for self-recovery or extrication of the gun or limber.

Entering production in 1935, the CD/SW was issued as a tractor for the 18 pdr, 18/25 pdr and 25 pdr gun-howitzer equipping Field Regiments of the 1st and 2nd Divisions of the BEF, consequently the majority of these vehicles were abandoned in France. Small quantities were sold to the armies of Australia, New Zealand and Eire, while a semi-armoured version appeared in 1937 with a 'droop-snoot' lightly armoured bonnet.

A light anti-aircraft tractor was produced in 1938 to tow the newly introduced 40 mm Bofors. The prototype vehicle had a body that appeared to owe much to the furniture removal pantechnicons of the period. A more military looking version replaced it in production and continued into the early 1940s. This version was still in service at the end of the War, although largely supplemented by the Bedford QL 4 × 4 and CMP 4 × 4 LAA tractors.

In 1938, encouraged by the successful development of 4 × 4 vehicles, a specification was formulated by the War Office for a field artillery tractor capable of handling the new 25 pdr gun-howitzer. The tractor had to have a short wheel base 4 × 4 chassis with winch, and a good cross-country performance with towed load. Accommodation was to be provided for a gun detachment of six. With room for 96 rounds of ammunition, gun tools, crew personal kit, etc., as well as room for the traversing platform, so characteristic of the 25 pdr, which was carried on brackets over the rear of the body; this could also be slung under the gun itself.

The first tractor to enter production was the Guy design, this was designated the 'Quad Ant' as the chassis was a quadruple-wheel-drive version of a 4 × 2 15 cwt infantry truck known as the 'Ant'. The name 'Quad' became the generic term for this family of hump-backed multi-facet bodied tractors that were to become such a familiar sight on every battlefield. The Quad Ant remained in production until it was superseded by a 15 cwt 4 × 4 GS truck version in early 1943.

The Guy model was followed in service by the Morris-Commercial C8 Mk I (1938), which itself had been preceded by some peculiarly bodied prototypes. The Mk I and Mk II (1939) versions had permanently coupled four-wheel drive with a five speed gearbox and 70 bhp 4 cylinder 3,519cc Morris EH engine.

The Mk III appeared in 1942 with minor body variations and optional front drive. It was followed in 1944 by a later Mk III with an entirely new body – the Type 5. This was of more conventional design with a canvas tilt. This version could also be used as a tractor for the 17 pdr anti-tank gun and remained in service until 1958, having seen action in Korea.

The Karrier Company produced a version of the basic FAT for the Indian Government, it used the basic Humber armoured car chassis, but with the engine moved to the front. This, the KT-4 was somewhat larger and more powerful than the Guy and Morris versions, and had a partially open canvas-covered crew compartment. About 400 were produced between September 1939 and January 1940 for the Indian Army, and many saw service with the Indian divisions of the Eighth Army.

Production of British Quads amounted to about 5,000 units and this fell far short of requirements. The Canadian industry came to the rescue with CMP versions of the Quad; Ford produced the FGT and Chevrolet the CGT tractors on their 101 in. wheel base 4 × 4 chassis. Body development followed British designs and ran to six versions, as follows:

| | | |
|---|---|---|
| *1* | FAT-1<br>Body 7A-1 | Similar to early British bodies with enclosed steel body and roof ventilators. Front end similar to No. 11 cab. |
| *2* | FAT-2<br>Body 7A-2 | As FAT-1 but with large canvas-covered roof-opening and rear ventilator flaps that could also be used as observation ports. Cab No. 12. |
| *3* | FAT-3<br>Body 7B-1 | First version to use No. 13 cab with body No. 7B-1. |
| *4* | FAT-4<br>Body 7B-2 | Production version of FAT-3 with No. 13 cab giving a 4-door body. Fitted with pneumatic tyres consequently a spare had to be carried, this fitted in place of traversing platform. |

*The Japanese and Russians favoured fully-tracked tractors, based on agricultural designs, for towing artillery. The Type 92A tractor came into service in 1932.*

General Patton

*Mack NO2 $7\frac{1}{2}$ ton $6 \times 6$ artillery tractor, in full US Army regalia! Note the bow at the rear from which a hoist lifted the gun trail on to its towing pintle. The NO2's powerhouse* (top right) *the Mack EY 6 cyl. 707 cu. in. engine, develops 159 bhp at 2100 rpm, while the Garwood 5 MB winch of the NO2* (bottom right) *has a capacity of 25,000 lb and is driven from a PTO. The yellow-painted connection is an air coupling.*

| | | |
|---|---|---|
| *5* | FAT-5 Body 7B-2 | This was a winterised version of the FAT-4 and was capable of operating down to −20°F (−29°C). |
| *6* | FAT-6 Body 7B-3 | The equivalent of the dual purpose British Type 5 body, capable of towing either the 25 pdr or 17 pdr anti-tank gun. |

Quantities of these chassis were supplied to Australia where locally designed and built FAT No. 8 and No. 9 bodies were fitted. These versions dispensed with the limber.

A total of 22,981 complete Quads and chassis were produced by the Canadian companies during the War period.

British field artillery tractors were, as a class, underpowered for the task of towing a 25 pdr and limber together with a load of ammunition, crew and stores – this resulted in a considerable use of the gears. The Guy Quad engine developed only 53 bhp, against the 70 bhp of the Morris, 85 bhp of the Chevrolet and the 95 bhp of the Ford V-8. Other problems were the tendency to blow cylinder-head gaskets, which often resulted in cracked blocks. Crown wheels and pinions also failed, as did front axle CV joints, and cracking spring hangers and engine sub-frames could not be readily repaired by welding.

The original vehicles were also unsuitable for desert use because of the completely closed-in crew compartment, and a canvas-covered roof aperture appeared in later models.

However, the Quads gave long and faithful service, and are still remembered with affection by the gunners who went to war in them, despite their uncomfortable ride and lack of protective armour. The Director Royal Artillery considered the field artillery tractor unsatisfactory, but despite new designs produced in the late War years, the Royal Artillery never found a Quad replacement, instead, in the post-war years, they utilised GS truck conversions such as the Bedford RL and Landrover.

Mention has already been made of the influence that German semi-track design had on British thinking, and the British vehicle it influenced warrants a more detailed study. The new semi-track British tractor was intended to serve as a tractor for the 25 pdr gun-howitzer, 17 pdr anti-tank gun and 40 mm Bofors LAA gun, and initial design work was carried out by Morris-Commercial. The final design and prototype production was handed over to Bedford, who had six models under test by December 1944. Known as the 'Traclat', a contraction of *Trac*ked *L*ight *A*rtillery *T*ractor', the Bedford BT was powered by twin Bedford 3.5 litre engines developing a total of 136 bhp and transmitting power, via two separate shafts, to a single clutch and five-speed gearbox which contained the differential track steering mechanism. Maximum speed was governed down to a maximum road speed of 30 mph but the Traclat was capable of maintaining this speed with a towed load up a 1 in 30 slope; whereas, the Morris C8 under similar circumstances would have been down to 3rd gear and 15 mph. The Traclat could also tow a 25 pdr up a 1 in 2 slope.

The track was an unashamed 'chinese-copy' of the ZgKw 5 design but with improvements intended to ease maintenance requirements and prolong track life. Track links were light steel castings with detachable rubber pads and track pins carried in sealed needle roller bearings. Pad mileage was approximately 2,000 miles, and the ultimate track life was that of the vehicle itself. Ground pressure, fully laden, was 8.1 lb per sq in. The vehicle was fully waterproofed and preparation for deep wading consisted of fitting air intake extensions carried as part of the normal equipment.

The all-steel body was divided into three separate units, each of which weighed less than 5 cwt and could be removed by the crew. The forward portion housed the driver and detachment commander. The centre portion carried ammunition and gun equipment with drop sides for easy access, and fittings varied depending on the artillery piece being towed. The crew compartment at the rear could accommodate five men with their personal equipment. A full length canvas tilt, incorporating two hip rings, protected these compartments from the elements. Finally, a 5 ton winch was installed. The Bedford BT had a radius of action of 200 miles. The finish of hostilities meant that the Traclat never entered production, but at least one of the prototypes was 'demobbed' and saw service with a logging company.

As a precaution against the failure of the

Traclat design, a design contract was awarded to Albion Motors for a 6 × 6 light artillery tractor which had to incorporate as many standard components from existing FT 11 3 ton 4 × 4 GS and BY5 3 ton 6 × 4 GS models as possible in order to reduce cost and development time, and facilitate early production. The Albion FT 15N was probably the first British low-silhouette design and had a 95 bhp engine, an 8 ton winch, and was capable of towing the 25pdr, 17pdr, Bofors or 6pdr gun, no limber being needed as all ammunition was stowed in the body.

*Morris Commercial C8 Mk III with field artillery tractor body. This was superseded in production by an open body with canvas tilt, the wheel of evolution had turned a full circle.*

Below *The Guy Quad-Ant tractor 4 × 4 field artillery was produced to a similar specification as the Morris 'Quad', but production terminated in 1943 to give priority to the 15 cwt GS version.*

*Typical CMP field artillery tractor. This is a Ford FGT fitted with the fourth type of body, carrying a spare wheel on the rear sloping deck indicating that run-flat types were not fitted.*

During tests at the WVEE, the FT 15N averaged 22.9 mph towing a 17 pdr round the Hindhead circuit. Although intended as a stop-gap design, the FT 15N and FT 15NW (wade-proofed) entered production and 150 were produced between September 1944 and the end of the War. The author 'spotted' an FT 15N, stripped of its body, with a jib added to the rear chassis and a Perkins Diesel fitted, in Hampshire in mid 1978.

An unusual prototype light artillery tractor was produced by Dennis in 1943, originally intended to be an 8 × 8 configuration with twin Bedford engines as in the BT, a radical redesign was undertaken and the prototype finally emerged as a 6 × 6 with a Leyland 9.8 litre engine and light armour protection, with a central driving position. The 'Octolat' (*Octo* (8 × 8) *L*ight *A*rtillery *T*ractor) had skid steering and no springs, the low pressure tyres absorbed suspension loads. No production was undertaken but the prototype survived for many years.

In the medium artillery tractor class the British had a good vehicle in the AEC Matador, this first entered service in 1938 and continued in RAF use until the early 1970s. The Matador was developed from original FWD designs, by way of a Hardy Motors 4/4 4 × 4 chassis which utilised many AEC components prior to being absorbed into the AEC group. Early versions were powered by a petrol engine and had a civilian-type cab. The body could accommodate a nine man gun crew, as well as ammunition and gun equipment, and was distinguished by a metal roof with side storm-curtains. In this form, it was used to tow such vintage artillery pieces as the 6 in. 26 cwt howitzer and the 60 pdr in rubber tyred (pre-pneumatic) form, as well as the more modern 3.7 in. AA and 4.5 or 5.5 in. guns. The diesel engined 0853 model made up the bulk of the 8,612 Matadors produced for the Army. 417 were used by the RAF. This diesel version could be distinguished by its high curved cab roof and canvas topped body, and it was powered by the very reliable 6 cylinder 0853 7.58 litre diesel, producing 95 bhp. The only major problem with the Matador was the winch which was prone to bearing-seizure, and that meant lifting the complete body before the winch itself could be removed.

Apart from its intended duty as a gun tractor, the Matador was pressed into service as a tank transporter in the Western Desert, towing Rogers trailers, often with the top half of the cab removed. At about the same time, a successful field modification was carried out at Tobruk to convert the Matador into a recovery vehicle. The body and top cab were removed and a CMP 30 cwt steel GS body fitted. A rigid jib was welded up from 'H' section-angle, and a winch constructed from other vehicle mechanical parts, including an Italian differential. The resulting vehicle had a 6 ft maximum height and a 5 ft 6 in. jib lift; its main advantage over the standard 6 × 4 3 ton Gantry breakdown vehicles was its more powerful engine, its four-wheel drive and low silhouette.

A basic Matador was converted into a semi-track tractor, the track design utilising many Valentine tank suspension components. This vehicle was intended as a 6 pdr or 17 pdr anti-tank tractor, but the weight of the track unit was excessive and as there was no great improvement in performance over the standard Matador, this vehicle did not warrant production.

Another experimental model was a tractor intended to tow the Twin 6 pdr AA equipment. The following additional equipment was fitted for this role:

*Early production version of the CD/SW LAA tractor with standard body. The enclosed prototype body having been completely redesigned to give a lower silhouette.*

*An early production AEC Matador (diesel) towing a 6 inch 26 cwt howitzer of 3rd RHA. Note lack of access door in body fitted to later vehicles.*

*1* 10 kva Alternator driven by V-belts from pulley at the rear end of the gearbox.
*2* Cable drum in body to lead power out to gun.
*3* Hand control in driver's cab for governed fuel pump.
*4* Stowage for 120 rounds of 6 pdr ammunition.
*5* Seating for seven crew in body, which was of steel roofed type.
*6* Warner electric and 2-line air-pressure connections fitted for gun brakes.

The Twin 6 pdr never entered production and very few tractors were produced.

The Park Royal body building concern produced a collapsible artillery tractor body for mounting on the Matador. This was an experimental airportable version of the standard MAT body designed to fold flat, it had a tubular superstructure similar to a GS truck.

The Matador was also fitted with a body similar in construction to the Workshop No. 4 Mk V and housed an Inert Gas Plant No. 1, intended to produce dry inert gas for use in charging gun recuperators.

Apart from these soft-skin roles, the Matador chassis was used to carry armoured command vehicle bodies and a 6 pdr anti-tank SP known as the 'Deacon'. It was also used as a basis for several marks of AEC heavy armoured cars.

The Canadian equivalent of the Matador was a hybrid vehicle consisting of the US built FWD SU-COE 5 ton 4 × 4 chassis which was fitted with a British MAT body similar to that of the Matador or a Canadian designed all-steel body, the 7K1. The SU-COE was used extensively by the Canadian artillery to tow the 5.5 in. gun or the 3.7 in. AA gun, and was used in smaller numbers by the Royal Artillery Medium Regiments. A feature of the FWD was that the front axle drive was permanently coupled, and in British service the fifth gear was blanked off.

Moving on to the heavy artillery tractor class, we again find that the British had a first class vehicle available. This was the Scammell 'Pioneer' R-100 6 × 4 which first entered service in the mid 1930s and remained virtually unchanged throughout its service life. The only drawback was that production never kept pace with demand.

Powered by a Gardner 6LW 8.4 litre 6 cylinder diesel developing 102 bhp, the Scammell was used as a tractor for the 60 pdr and 6 in. guns during its early career, it then coped with the heavier 8 in. gun and 7.2 in. howitzer. An overhead 10 cwt block and tackle was installed to load and unload ammunition and, on occasions, to lift gun trails on to the towing hook. There was seating for a nine man gun detachment in the body, together with room for ammunition and equipment. Either single or two-line air brake connections were fitted to cater for the different types of gun brakes. The Royal Engineers also used a Scammell whenever they could, for towing 18 to 20 ton class trailers carrying heavy engineer plant.

In an attempt to meet the need for heavy tractors, Albion were asked to design a complementary vehicle to the Scammell. In prototype form the resultant vehicle, the CX 22, had a cab reminiscent of the Matador but in production it reverted to typical Albion style, on a chassis similar to the CX 23N 10 ton GS, and fitted with the standard Albion EN 244 9 litre 6 cylinder diesel engine, developing 100 bhp. Production commenced in November 1943 and 532 vehicles were completed by June 1945 when the contracts were terminated. The cab could be split at the waist line and the body dismantled for shipping. A Scammell 8 ton vertical winch was fitted but no overhead gantry, loading was by means of skids.

The CX 22 duplicated the role of the Scammell and was its equal in performance.

In order to cope with the heavier US artillery pieces about to be supplied to the British Army, a new tractor was developed from the AEC 0854 6 × 6 chassis. This low-silhouette vehicle was powered by the AEC 9.5 litre 6 cylinder diesel engine of 150 bhp that originally saw service in the Valentine tank. It could carry a payload of $4\frac{1}{2}$ tons and tow either the 7.2 in. howitzer on the US Mk 1 carriage or the US 155 mm howitzer weighing 14 tons.

A second prototype, produced in September 1944, had an armoured cab and crew compartment, but was otherwise similar to the original model. It was assigned for a 'special role'. Could this vehicle have been intended to tow the giant 32 pdr anti-tank gun then under development and weighing over 10 tons?

These prototypes eventually developed into the 10 ton family of post-war FV 11000 Series AEC vehicles.

Although the medium range of tractors also doubled as prime movers for the 3.7 AA gun, a special design was produced to tow the 40 mm Bofors light AA gun. As already mentioned the first of these was the Morris 6 × 4 CD/SW to be followed by the Bedford QL LAA tractor.

The Bedford QLB 4 × 4 Tractor L A A utilised the basic QL chassis fitted with a 5 ton winch.

*A Matador with equipment and ammunition stowed for 4.5 inch role. The boxes under the rear-facing seats contain fuses, propellant charges are carried in metal boxes in the centre of the body.*

The body was basically in two sections, the forward crew cabin accommodated five of the nine man crew with stowage for spare gun-sight, pioneer tools and crew kit. The spare gun barrel locker, encroached into the cabin from the rear body. This rear body, carried a stack of lockers at each side over the rear wheels, each housing four ammunition boxes, two stowage boxes topped by two open lockers. The spare-barrel box ran down the right-hand side into the crew cabin. Two more crew seats were fixed to the top of the box and the gun spare wheel was attached to the rear of the crew cabin, the whole could be covered by a small tilt.

The Canadians produced two main types of LAA tractors. The first was based on the 134 in.

*The 6 pdr A/Tk gun tractor version of the Guy Quad-Ant 4 × 4 15 cwt can be distinguished by the winch rope guide rollers on the front bumper and the translucent 'windows' in the tilt.*

wheelbase 30 cwt chassis uprated to 3 tons by fitting stronger springs and larger tyres. A specially compartmented body housed much the same equipment as the QL but stowage lockers were accessible from outside the vehicle. The later style was also produced by Ford and Chevrolet but was mounted on the 30 cwt 4 × 4 chassis with No. 13 cab, the body being similar.

A small number of US 6 × 6 $2\frac{1}{2}$ ton trucks were pressed into service by the British as Bofors tractors, but no special stowages were fitted. Ammunition was stowed at the forward end of the body with a spare barrel down the middle, the crew used the standard hinged slatted seats. When a predictor accompanied a Bofors battery, this was carried in an instrument truck, usually a GMC AFKX-352, which also towed a two wheeled ammunition trailer.

Tractors had also to be provided for the various anti-tank guns that equipped the British Army between 1939 and 1945.

The first of these guns was the diminutive 2 pdr and was initially manned by the Royal Artillery who used various marks of Dragons as towing vehicles. However, when the 2 pdr was issued to infantry battalions a cheaper and simpler tractor had to be found. The solution lay in adapting the 15 cwt 4 × 2 infantry truck which in 1937 was entering service in quantity. The gun could either be towed, or loaded into the body via light ramps, and porteed into action, but not fired. A slightly higher tilt was fitted and the 15 cwt 2 pdr portee could be distinguished by the translucent side panel in the tilt that served as a window. Bedford, Morris and Guy all produced 2 pdr tractor/portees.

A more complicated portee body was introduced to allow the gun to be fired forward over the bonnet. The body was fitted with hinged ramps and curved rails to locate the gun wheels, which were then clamped securely into the firing position. The whole body could be tilted upwards at the front, about hinges at the rear, to facilitate loading which was accomplished by a small hand winch. This type of portee body was fitted to the Morris C8/MG 4 × 4 15 cwt and CMP 4 × 4 30 cwt CGT chassis. When the 2 pdr was declared obsolete, the surviving portees were converted into tractors for the 17 pdr anti-tank gun. The disadvantage of this portee was its limited traverse, and forward firing meant that the driver had to dismount. Locally designed and built portees were evolved in North Africa using CMP chassis with open cab and rudimentary body. The 2 pdr was mounted facing to the rear over the back axle giving it a good field of fire but was very exposed and suffered severe casualties in action. Despite this, a 6 pdr version came into service once this gun was available in sufficient quantities.

The 6 pdr was generally towed by either the Universal, Loyd or Windsor tracked carriers but a special 6 pdr tractor version of the Guy Quad-Ant 15 cwt 4 × 4 GS was produced in early 1944. A winch was fitted as standard and there were four seats in the body with stowage for ammunition. Twelve prototypes were produced and issued for user trials.

A more sophisticated 6 pdr portee than the desert version was produced in Britain and mounted on open cab versions of the Bedford QL, Austin K5 and CMP 3 ton 4 × 4 chassis. These saw service in Tunisia and in the closing stages of the desert campaign but their tactical usefulness was over, and many were converted into 3 ton load carriers.

The entry of the 17 pdr into service with towed batteries of Royal Artillery anti-tank regiments meant that a new generation of tractors had to be produced. As all too often happened, no suitable

*The Albion FT15N was intended as a stopgap field artillery tractor pending introduction of the Bedford BT 'Traclat'. However only six 'Traclat' prototypes were produced while the Albion saw limited service use.*

vehicle was immediately available and conversions of existing trucks had to be made. Surplus Morris C8/MGs and C8/Ps were among the first to be converted together with the CMP versions. The Morris C8/P was a self-propelled predictor, built to accompany SP Bofors batteries, but rendered superfluous with the introduction of radar. These conversions had to seat a crew of eight together with 30 rounds of 17 pdr ammunition and equipment, resulting in what must have been one of the most overcrowded, uncomfortable and underpowered vehicles of all time. As the 17 pdr weighed 6,500 lb, compared to the 2,471 lb of the 6 pdr, and the 1,848 lb of the 2 pdr, more powerful tractors had to be found. Field artillery tractors with the Type 5 body were used, as well as simple conversions of the Ford WOT-8 30 cwt 4 × 4 GS truck together with M5 half-tracks. Later, turretless versions of the Sherman and Ram tanks were pressed into service, and a special tractor, designed around the Crusader chassis, was issued in time for the closing stages of the NW Europe campaign. Incidentally, more towed 17 pdrs were disabled by road accidents in NW Europe than by enemy action.

The British probably had fewer self-propelled light anti-aircraft guns mounted on modified truck chassis than the other combatants. The Bofors was fitted to a lengthened Morris C8 FAT Mk III chassis, which received the type number C9/B, and was supplemented by a similar installation on a modified CMP Ford F60B chassis. The gun had powered traverse and elevation, and could either be fired, while still, with wheels on the ground or on integral jacks in a semi-static role. The rear springs locked automatically when the gun was brought into action, to prevent 'bouncing' under recoil. In the later stages of the War, when the Allies had air

*Extemporised A/Tk gun portees were produced during the N African Campaign leading to more sophisticated designs. This CMP Chevrolet CGT was fitted with a British designed body for the 2 pdr A/Tk gun. Shown in the loading position.*

supremacy, these guns were used in a ground role rather than against aircraft and had a reasonable armour piercing capability.

In the lighter AA class were the single 20 mm cannon mounted on 15 cwt chassis. Polsten and Hispano guns had been available in a towed version, and the carriage was adapted to the self-propelled role simply by clamping the carriage to the chassis of the 15 cwt after removing the gun wheels. Ammunition stowage boxes were provided as was a guard rail to prevent enthusiastic gunners shooting off the unfortunate driver's head.

The Bedford MWC, Ford WOT-2 and CMP Chevrolet C15 and Ford F15 chassis were converted into light AA trucks, and for a time, equipped a platoon of the medium machine-gun battalions. Some CMP versions took part in the invasion forces but were soon withdrawn, the surplus cannon being fitted to LVTs for the Rhine crossing.

A large proportion of the German field artillery was horsedrawn throughout the War, but there were some very fine tractors available from the extensive range of semi-track vehicles. Probably the best known was the 8 ton Sd Kfz 7 medium semi-track, produced by Krauss-Maffei as the KM m 11, which was a prime mover for the 88 mm Flak. At the other end of the scale was the equally well known Krupp L2 H43 Protzkraftwagen used to tow 2 cm light anti-tank and AA guns.

The 15 cm heavy artillery pieces were normally towed by semi-tracks in ZgKw 12 class and the heaviest 18 ton chassis were used to mount SP 88 mm Flak guns.

The Germans also used the Steyr RSO/01 as a gun tractor and had other fully tracked tractors available in the models T3 and T9 from the

*With the demise of the 2 pdr A/Tk gun many portees were converted to 17 pdr A/Tk gun tractors. This is a converted Chevrolet CGT.*

*The Mack No. 2 $7\frac{1}{2}$ ton 6 × 6 artillery tractor. Designed to tow heavy artillery such as the 155 mm gun and 8 inch howitzer. This particular vehicle has the special trail clamp fitted to which trails of certain guns were attached.*

occupied Czech factory of Praga.

There were some heavy wheeled tractors in service, these stemmed from the civilian multiple-trailer prime movers that were a familiar sight on the pre-war autobahns. The extra large wheeled tractors, usually with chevron patterned steel wheels, used mainly on the Eastern front, were a hangover from World War I German designs.

The German drawbar couplings, with automatic locking catches, are worthy of closer attention, as they were of particularly sophisticated design, probably due to their use in commercial trailer 'road trains' (multiple trailers), which were closely legislated and which were required to be capable of being safely operated by one man.

The heavy drawbar coupling fitted to the 18 ton Famo was capable of absorbing a pull of 12 tons. The securing pin was a close fit in the gun trail eye and was mounted in a universal coupling, so that easy alignment could be achieved by one man. Shock absorbing was achieved by a series of conical steel washers, similar to the Belleville washers used in some aircraft gun-recoil systems, and very like the German railway-buffer design. Very little snatch or kick-back was evident with this type of coupling.

This was in strong contrast to British designs which apparently horrified the Germans with their crudity.

The standard British drawbar eye was a round section, fitting into a flared jaw fixed rigidly to a rear frame cross member and secured by a loosely fitting pin, with no shock absorbing system. An improved design consisted of a lateral leaf spring attached to the chassis with a central hook incorporating a spring-loaded retainer, however, because of the large hook-to-eye clearances, there was a lot of inherent snatching with this design. A fact which was brought home by a spate of failures with the 3.7 in. AA gun carriage towing eye.

The Italian artillery tractors were interesting, in that they reflected the problems faced in manoeuvring guns in mountainous terrain. They featured a short wheel base, large diameter wheels and steering on front and rear axles to reduce turning circles. The unique Pavesi articulated chassis was incorporated in the Fiat/Spa Pavesi P4-110 heavy artillery tractor which saw service in relatively large numbers. All of these tractors suffered from lack of stowage space and payload, due to dimensional design limitations.

A large range of fully tracked agricultural tractors had been put into production by the Russians in the early 1930s, and these designs were taken into military service in the Red Army as artillery tractors for medium and heavy guns. The Red Army also had fully tracked heavy tractors but these were equipped with enclosed cabs and cargo-type bodies.

Many lend-lease vehicles were supplied to the USSR by the USA and Britain and Canada, some of these vehicles such as the Studebaker $2\frac{1}{2}$ ton 6 × 4 and British Ford WOT-8 3 ton 4 × 4 were fitted with a multi-barrelled 'Katuscha' rocket launcher, a weapon christened the 'Stalin Organ'. The influence of these imported vehicles on indigenous designs was considerable and manifested itself in the post-war products of the Russian automotive industry.

As a matter of expediency, most nations pressed cargo trucks into service as gun tractors. The Jeep is a good example, it was used by the US and British airborne artillery regiments to tow the 75 mm M1 howitzer and airborne version of the 6 pdr anti-tank gun. It was also used by the British in the mountain divisional artillery for the 3.7 in. pack howitzer, while the Australians used it in New Guinea to tow the 'Baby 25 pdr'; which was almost unrecognisable as a 25 pounder with its short barrel, small wheels, and narrow track. The 'Baby 25 pdr' lacked the familiar square shield but was well able to negotiate narrow jungle and mountain tracks behind its diminutive tractor.

The vehicles already described have been in special purpose categories but the wide variety of designs warranted separate treatment, that is not to say that the other specialist vehicles were in any way less important or interesting in concept.

# AMBULANCES AND ASSOCIATED MEDICAL VEHICLES

The medical, signals, engineers, logistic and other supporting services, all had their particular requirements for specialist vehicles, to enable them to carry out their role as part of an integral mechanised fighting force.

Any military medical service needs to provide fleets of ambulances to evacuate wounded, mobile hospitals and laboratories. At the front line, a small vehicle with cross-country agility was needed and, when employed with armoured formations, this vehicle itself was often armoured (but always unarmed). Generally speaking, this front line ambulance would accommodate two stretcher cases or four 'walking wounded'. The Jeep was universally adopted for this purpose and many ingenious modifications were carried out to enable up to four stretchers to be carried. A simple framework was designed by the Canadian Army to carry three stretchers on the Jeep, and this was adopted by the British Army as standard, all Jeeps received by the British from the end of 1943 had sockets to take this framework welded in place before issue.

The Jeep was, however, not protected from the elements and the Humber 4 × 4 light ambulance had good cross-country performance and could accommodate two stretcher cases in greater comfort.

The same chassis was used for the Humber light reconnaissance car Mk III. By far the most numerous British ambulance was the Austin K2 4 × 2 heavy ambulance. The body was of simple construction, a wood frame covered with leather-cloth, well insulated, heated and accommodating

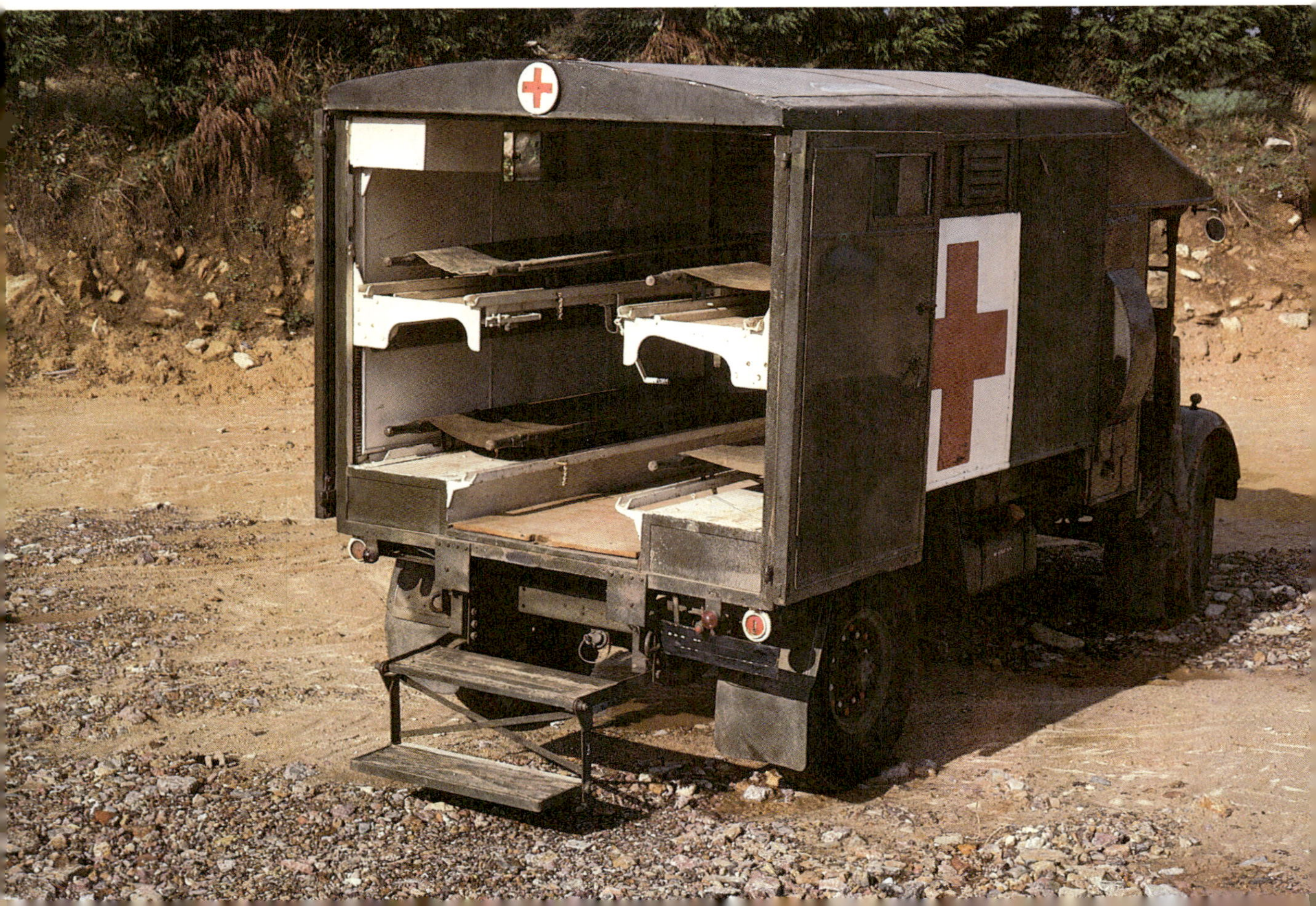

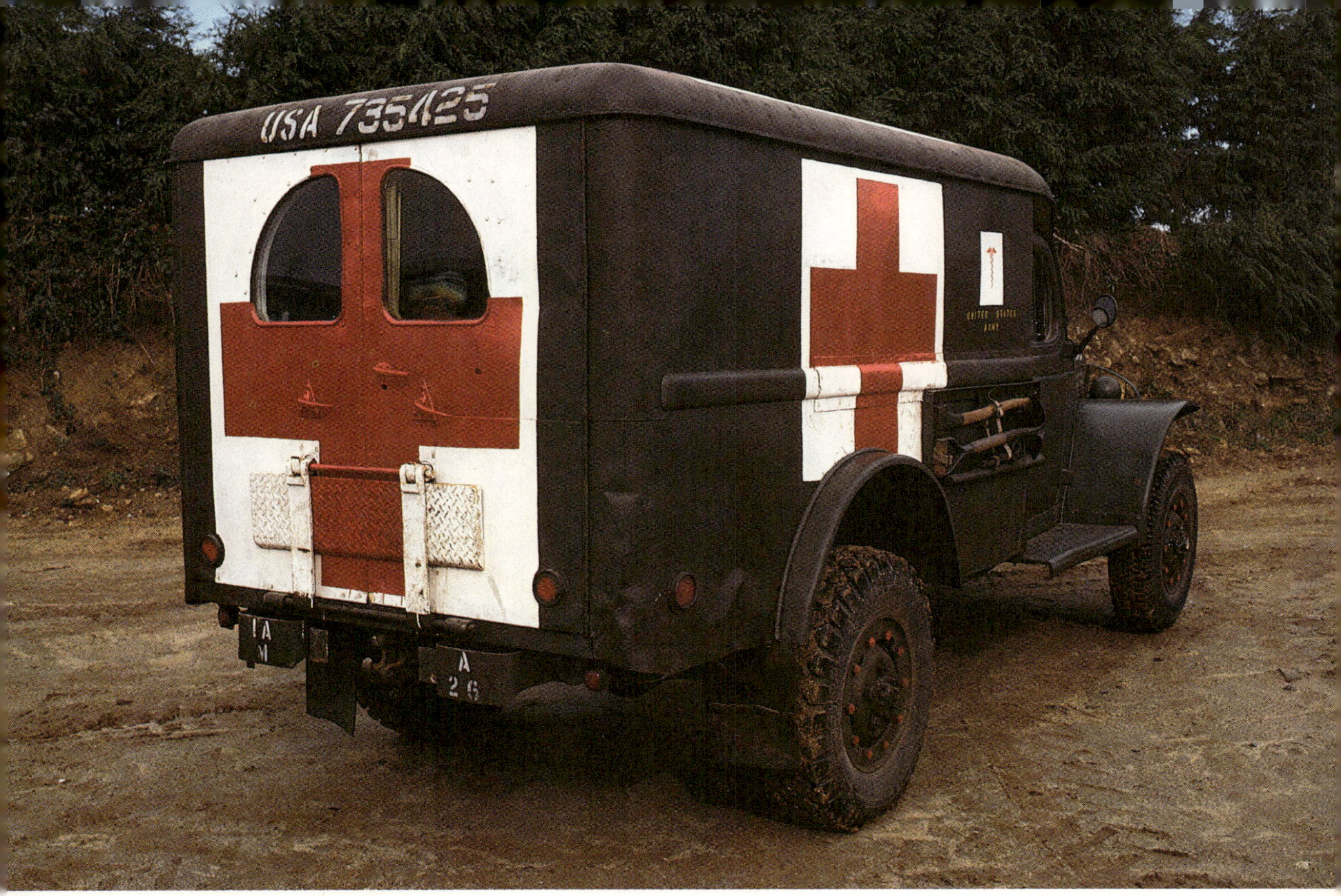


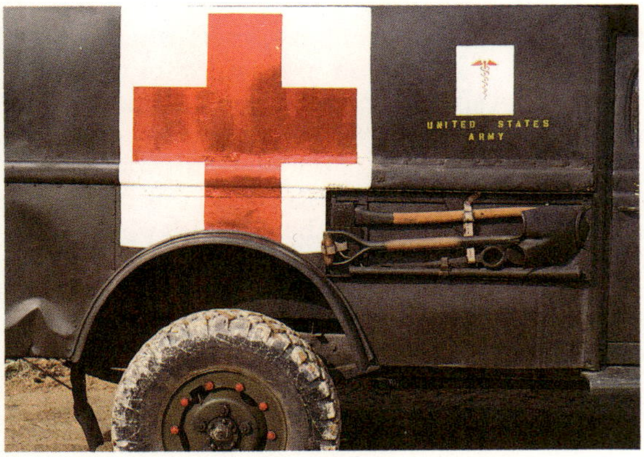


Top Left *Austin K2/Y heavy ambulance 4 × 2. Affectionately known as the 'Katie'. Over 13,000 were produced for the Allies but not before some initial teething troubles were overcome.*

Bottom Left *The K2 showing the 4 stretcher layout in the wood-framed leathercloth-covered body. The upper stretchers could be lowered to provide seats for walking wounded.*

Top Right *The American equivalent of the Austin K2, the Dodge T214-WC54 had the advantage of all-wheel drive and steel body.*

Bottom Right *Dodge T214 details of right-hand side showing standard US Army pioneer tool-tray stowage and Medical Corps insignia.*

*Many basically civilian designs were hastily converted for military purposes at the outbreak of war. Morris TMV chassis were equipped with light-ambulance bodies for use in non-operational areas.*

four stretcher or eight sitting cases, or combinations of both, plus a medical attendant.

This body was the result of many years of pre-war development by the Royal Army Medical Corps. Produced by Mann Egerton, the specialist luxury car body builders, the No. 2 Mk I/L ambulance body was also fitted to the Bedford ML and Morris CS11/30F chassis.

Many impressed civilian vehicles were fitted with ambulance bodies of simple design, but these were retained for home use. The K2 was by far the most numerous of the various types of British ambulances and 13,102 were produced, many going into RAF service or to the American forces in Britain. The British Army had over 15,300 ambulances of all types in service at the end of the War.

Of the armoured vehicles used as ambulances, probably the most unlikely was the Daimler 'Dingo' scout car. This was issued to armoured regiment medical officers as a battlefield transport, but often carried a stretcher on its engine deck. The Universal carrier, White scout car, an armoured version of the CMP 15 cwt CI5TA and the M5 half-track were all pressed into service as ambulances. The White scout car was unsuccessful, as stretchers had to be lifted over the sides but the M5, with its rear door, was easier to load and a

*A Willys 'Jeep' fitted with British-designed two stretcher modification. An extended canvas hood was fitted to protect the casualties. Superseded by three-stretcher layout.*

special modification devised by the British involved moving the left-hand side fuel tank to the rear of the right-hand tank. A collapsible framework was then fitted into the body, this was capable of supporting four stretchers or, if only half extended, two stretchers, and normal seating was then available on one side for walking wounded. A special armoured ambulance version of the Canadian CMP C15TA was entering production at the end of the War, this was similar to the earlier armoured GS truck, but had sides and rear extended to cab height with permanent fittings for stretchers and medical equipment.

There were other vehicles associated with the medical services, these were generally attached to the field hospitals to which the ambulances delivered their unfortunate passengers.

The mobile dental surgery was a typical example, a house-type body, mounted on the standard Bedford QL chassis, contained special laboratory equipment associated with dental surgery. The Bedford towed a two-wheel caravan trailer, in which was housed a small office and an operating area with dental chair and all the normal dental equipment. This vehicle visited static army units, to carry out routine dental work; more importantly, it also visited battle areas to assist in the temporary rebuilding of shattered jaws, prior to the casualty being evacuated to a base hospital for more extensive treatment.

Special refrigerated bodies were fitted to

15 cwt chassis for the transportation of blood plasma, and laboratories such as those on the Austin K6 6 × 4 3 ton chassis matched blood samples, checked water and analysed suspect food, poisons, etc., as well as identifying bacteria. Apart from an integral water system, built-in sinks, worktops and refrigerator, it was equipped with microscopes, balances, autoclave, centrifuge and fume cupboard, together with storage for the various chemicals necessary to carry out the various tasks involved.

Another essential facility to be provided was that of X-ray services. First developed in the early 1930s and fitted to 6 × 4 3 tonners of the period, the X-ray house-type was later carried by the Bedford OY 4 × 2 3 ton and CMP Ford 6 × 4 3 ton. This 14 ft body contained up-to-date X-ray equipment and the means of developing film. Electric power was supplied by a petrol-driven generator, which was housed in the forward part of the body but which could be removed and operated outside the vehicle. Thus, X-ray examination of casualties' wounds and fractures could be provided by one vehicle at the forward area Casualty Clearing stations.

Each battalion, or equivalent unit of the British Army, had an RAMC doctor (the Medical Officer) and medical staff as well as the regimental medical orderlies, 'the medics'. In order to provide the MO with mobile medical facilities, it was usual to equip a 3 ton GS lorry with basic medical equipment to act as a Regimental Aid Post. This conversion usually consisted of a special wide step fitted to the tailgate, easily removable fittings for two spring supported stretchers, a bench for four sitting patients, stowage bins across the front for medical stores as well as the regimental issue 'medical hamper', table etc. Side tents or penthouses were 'optional extras', dependent on unit resources and resourcefulness.

Vehicles fulfilling similar functions formed part of the inventory of most armies. In addition, both Germany and the USA used buses in the ambulance role. These were generally standard bus bodies, but with large rear doors, removable seats and stanchion fittings to support stretchers, and were mainly used outside combat zones to transport patients.

The standard US ambulances were based on civilian panel van designs, and went into production as the Dodge $\frac{1}{2}$ ton 4 × 4 ambulance with accommodation for four stretchers or eight sitting patients. This type was superseded in production, in 1942, by the larger $\frac{3}{4}$ ton model with similar internal arrangements, both models were supplied in small numbers to the British Army under lend-lease.

Laboratory bodies were fitted to the standard $2\frac{1}{2}$ ton 6 × 6 GMC chassis.

In German military service, all ambulances bore the designation Kfz 31 regardless of the type of body or the chassis to which it was fitted. Typical examples were the Steyr 640/643 6 × 4, the Phanomen Granit 25H and the Auto-Union Horch 4 × 4; these served in large numbers together with smaller quantities of captured and impressed vehicles from occupied territories.

# COMMUNICATION VEHICLES

Communications is a vital military support service, and wireless or radio contact, within mobile formations down to individual AFVs, is essential if battle control is to be maintained by the commanders. Artillery fire has to be controlled and corrected from forward observation posts, and logistical support must be arranged. Once the advancing formations have moved on, lines of communication troops must set up bases and supply-dumps etc., and more permanent signals systems must be established in the form of telephone and teleprinter services. A range of

specialist vehicles is required to erect, maintain and operate these services (the workshop trucks have already been described).

At the outbreak of hostilities, the British Army had a tank force that was well equipped with radio sets, although some light armoured cars had to dispense with part of their armament in order to accommodate a wireless set – crews were given a choice between the Boys anti-tank rifle or a No. 11 wireless set, the latter being more popular. Brigade and Divisional HQs were the first to have armoured command vehicles (ACVs) on strength to provide the commanders with radio links forward to the fighting units and rearwards to the next senior formation.

As there were never enough ACVs to satisfy all demands, infantry formations normally had soft-skin radio trucks. These included the normal GS trucks equipped to carry radios and with suppressed electrical systems etc. being identified by the suffix FFW (fitted for wireless). The ubiquitous Jeep, the 8 cwt and 15 cwt trucks of British and CMP manufacture, together with 8 cwt heavy utilities were all used in varying numbers for the FFW role and were usually fitted with either the No. 9, 11, 19 or 22 W/T sets. A small petrol-driven generator set called the

Top Left *The White M3A1 car scout 4 × 4. Although it had limited armour protection and difficult access, it was popular as a command vehicle, armoured ambulance, fitters vehicle, etc. Over 20,000 were produced and many were supplied under lend-lease.*

Top Right *The Chevrolet C8A truck heavy utility 4 × 4 entered production in 1942 as part of the vast range of Canadian military-pattern vehicles. The IC2 wireless body had accommodation for a No. 9 or 19 W/T set, 3 operators and a charging set.*

Bottom Right *The GMC Model CCKW-353 truck $2\frac{1}{2}$ ton 6 × 6 cargo was the workhorse of the US and many Allied Armies. The basic chassis was used for the 'Duck' and also mounted specialist bodies.*

W/T
Z4236751
40
550 CPC

USA 4800572
TRK - 33

'Chore Horse' was carried to charge the radio batteries, as the vehicle engine could not be expected to run for long periods while stationary. These vehicles were issued to forward formations and their supporting arms, such as engineers, artillery and quartermaster units, and were manned by unit signallers for communication links to Brigade level.

At Brigade HQ level, the Royal Corps of Signals took responsibility for manning communications and the equipment became more sophisticated. The 13 ft house-type body, fitted to either the Bedford QL or CMP 3 ton 4 × 4, carried a variety of telecommunications equipment and the positioning of doors, windows, and interior partitions could be varied to suit any particular installation. Normally divided into two compartments, the typical layout of a 'Command, High Power vehicle' would be a command cell forward, with seating for five, map boards, tables, and headsets and remote controls for the radios which were situated in the rear; the radios consisting of a No. 53 set, 19 set and R107-type receiver with three operators. Other vehicles were equipped for long-range radio monitoring of friendly radio traffic to check procedures and security, they also listened to enemy stations to gather intelligence information.

The need for a more permanent communications system in a static defence or rear area, meant that new telephone lines had to be provided, as any civilian network had usually been destroyed. The British Army did not employ specially fitted vehicles for the task of erecting telephone lines but had a power-driven cable layer that could be fitted into a GS truck. The paid out cable was guided into position by a signaller with a long pole. As the War progressed, signal construction vehicles were obtained from the US and Canada. Their design was based on North American telephone company vehicles, and they carried a telescopic jib for lifting telephone poles, extension ladders, a winch driven from a PTO, cable dispensing gear and the special tools and spares required to erect telephone lines. This type of body was mounted on either the CMP Chevrolet 134 in. wb 4 × 4 chassis or the US 'K-43 Telephone Construction and Maintenance 1½ ton 4 × 4 GMC'. Telephone poles were transported either in a special two-wheeler trailer or in a normal GS 3 tonner and holes for the poles were drilled by a special power auger. This equipment was mounted on a standard 3 ton body, the auger, driven by a separate petrol engine, could bore a 20 in. diameter hole to a depth of 7 ft in 2 minutes in average ground. To complete the network, mobile telephone switchboards had to be provided, and again, these were contained in house-type bodies fitted with a variety of equipment, normally two 20 line switchboards, but up to 200 lines manned by five operators were available for Army HQ communications. These were termed Terminal Equipment Vehicles (TEV) and, in certain instances, a teleprinter was also installed at the expense of some telephone line facilities. Vehicles could be grouped together and linked to provide increased traffic handling capability. Power was supplied by trailer generators.

Like the British, the US Army Signal Corps had many standard vehicles fitted out to accept radio sets for combat unit communications work. Apart from the Jeep, the main types were the ½ ton 4 × 4 Radio and later the ¾ ton 4 × 4 Command, the Carryall and Panel trucks were also used but were limited standard vehicles. All Signal Corps special vehicles were given a K designation, an example being the 'K-50 Truck, Light Maintenance and Installation, ¾ ton 4 × 4', this being based on the Dodge ¾ ton weapons carrier chassis with a special body to house ladders and telephone line repair equipment. The K-44 was the standard earth borer (auger) and polesetter while the K-43, which was also supplied to the British, was a 1½ ton 4 × 4 Chevrolet chassis carrying a telephone construction and maintenance body. The Chevrolet 1½ ton 4 × 4 Panel delivery truck was the K-51 in Signal Corps service and there is photographic evidence that it was used by the British in Italy as a radio van.

The GMC 2½ ton 6 × 6 with standard house-type body, was fitted out with a variety of radio communications equipment, and was known as the K-53. The Americans also made greater use of trailers for signal purposes than did Britain or Germany, and these ranged from the K-38 ¼ ton trailer with cable splicing equipment and the K-37 cable reel and pole trailer to the 13 ton semi-trailer K-78.

*A Jeep, again on strength of 6th Armoured Division, fitted with radio, probably a local unit modification.*

The Wehrmacht telephone communications vehicles were normally based on the light 6 × 4 and 6 × 6 truck, with closed bodies in the Kfz 61 series providing mobile telephone exchanges; as did the Kfz 305/8 which was based on the Opel Blitz. The telephone line erection and maintenance vehicles were normally taken from the medium and heavy series vehicles, such as the closed body Kfz 72 Büssing Nag dating from the early 1930s. These carried poles, cross trees, ladders and cable plus all the winch line-laying equipment. The Büssing Nag Kfz 72 and 72/1 were also used to house teleprinters and towed a two-wheel generator trailer to provide power.

Various type of wireless equipment – for normal army communications, interception of enemy signals, radio location and transmission of spurious messages to confuse opposing forces, were installed in the standard Kfz 72 and Opel Blitz Kfz 305/15 to /27 variants. Other vehicles were fitted specifically to carry and erect wireless aerial arrays.

Like the Allied armies, the Germans had light vehicles fitted with wireless communication gear for their front line units, typically the Kfz 1 Volkswagen 82 light passenger car and the Kfz 2 Stoewer 40, whilst the Horsch 830 in the medium Kfz 15 class was used as a signalling service or as a telephone communication car. The Kfz 17 and 17/1 medium 4 × 4 with closed body on the Auto Union/Horsch and Opel EFm chassis were also used as radio cars, corresponding to the British heavy utility class.

# BRIDGING VEHICLES

In various army organisations the Engineer Corps were responsible for the construction of bridges and rafts, but the transportation of the actual bridging equipment, pontoons etc. varied from country to country. In the armies of the British Empire this task was performed by Bridge Companies of the Service Corps, accompanied by a Royal Engineer detachment. Until the invention of the Bailey Bridge system there were several different types of bridging equipment in service; these were Pontoon, Trestle and Sliding Bay, Folding Boat Equipment and Small Box Girder; all required special vehicles for their transportation. Also used, for the assault crossing of rivers and inland waters, were the inflatable rubber two-man reconnaissance boat, the collapsible wood and canvas assault boat Mk 3, the shallow draft wooden 20 ft storm boat Mk 1 with 50 hp outboard motor, and the Kapok assault bridge, which consisted of floats and 6 ft 6 in. decking sections which could build up to span 150 ft over still water.

This equipment was carried in a normal GS 3 ton 4 × 4 lorry which could accommodate either twelve assault boats, three storm boats or 70 ft of Kapok bridging material.

The storm boat could carry a Jeep, and rafts constructed from assault boats were capable of supporting a Jeep and 6 pdr anti-tank gun.

The special bridging bodies were fitted to standard 3 ton 6 × 4 chassis with minor modifications in certain cases.

The pontoon body could carry two pontoon sections, either centre or bow, one above the other. The first pontoon would be pushed into position over rollers set in the body floor and then hoisted up by four tower hoists with hand winches, when at maximum height, a second pontoon would be pushed underneath and the first lowered on top of the second and both lashed down. This body was fitted to the Albion BY series and Leyland Retriever. The Canadians fitted it to the special 201 in. wb Diamond 'T' 6 × 6. The Australians managed without a special vehicle, instead they utilised a flat bed semi-trailer with a CMP tractor. A pontoon platoon had an establishment of 24 'Lorries, 6 × 4, Pontoon', divided into four sections, with an HQ section consisting of a Jeep, 15 cwt, three 3 ton GS lorries, six motor cycles and a light trailer, two pontoon lorries were kept in reserve.

*This was the final version of the Retriever with full screen, revised 'Thornycroft' type radiator, and raised cab sides. The body was to transport trackway-bridging equipment, and the front bumper was part of the bridge-laying gear.*

The trestle and sliding bay body, fitted to the Albion BY 3 AEC Marshall and 201 in. wb Diamond 'T' was designed to carry components of the Folding Boat Equipment bridging equipment and came into service in the late 1930s. The long members were carried on sloping steel bearers and other parts were carried in the small forward portion or between the bogie arches.

Folding Boat Equipment came in two versions. The Mk II was introduced into service in about 1930, and components could be used to construct rafts or floating bridges up to Class 5. Pre-war 6 × 4 3 ton chassis were fitted with a sloping wooden framework to carry transoms, while the decking sections were stowed under this framework on a flat platform.

Three folding boats were transported on a special trailer towed by the FBE lorry. This equipment remained in service throughout World War 2 although it was declared obsolete in 1944, and had been replaced officially by the heavier FBE Mk III in 1939.

The Mk III boat was the same size as the Mk II but they were not interchangeable because of their different road bearer fittings. This equipment, like its predecessors, could be constructed as a raft or as a complete bridge in Class 9. The

*An Albion BY5 with folding bridge equipment body in difficulties crossing the river Garagliano in Italy. A Diamond 'T' 6 ton wrecker has come to the rescue.*

FBE Mk III body was entirely different from the Mk II and no trailer was necessary. Three folding boats were hoisted on bearers by four winch towers, and six road bearers were stowed underneath, together with other smaller bridge components. A locker fitting between the front right-hand winch tower and the front buffer cowl was used to stow rope etc. In order to reduce the loaded length of the vehicle, the cab was cut away to accommodate the buffer cowl and on British vehicles the fuel tank and spare wheel, normally positioned behind the cab bulkhead, were moved to either side of the chassis. The Mk III body was fitted to the Albion BY 3 and BY 5 3 ton 6 × 4 chassis and on the special Canadian 201 in. wheelbase Diamond 'T' Model 975. The Australians used a long flat platform semi-trailer with a simple tubular framework to carry three folding boats.

An FBE Mk III platoon consisted of an HQ and four sections, with a total complement of 16 FBE Mk III lorries (+ three spare) nine 15 cwts, three 3 tonners, six motor cycles and a Jeep. The platoon could carry a total bridge length of 560 ft with 300 ft of Somerfield tracking for approach roads. The bridge at Rees, spanning the Rhine, was approximately 1,300 ft long and would have required the equipment from three FBE platoons to construct it.

The fourth type of equipment to require special vehicle transportation was the Small Box Girder. The SBG Mk I, introduced in 1928, was quickly superseded by the Mk II. The final version, the Mk III, came into service in 1939. This bridge consisted of two main sections, the tapered end pieces, or hornbeams, and the symmetrical centre section, with deck panels and ribands. The SBG body was probably the simplest of the special bridging bodies. It consisted of a sloping steel framework with three transverse supports and an open stowage section behind the cab with small tool lockers on either side. Three SBG lorries carried the major components for a complete bridge, with a 3 ton GS lorry transporting the deck panels and ribands. SBG bodies were fitted to the AEC Marshall, Albion BY 1 and Karrier CK 6 British 6 × 4 3 ton chassis. The Canadians used the Diamond 'T', and once again, the Australians loaded everything on to a semi-trailer, the SBG Mk III (Aust).

This equipment was used in the early part of World War 2 and was later modified for fitment to the Churchill SBG carrier, which was operational with 79th Armoured Division from D-Day onwards.

With the invention of the Bailey bridge there was no further need for special vehicles to transport the basic bridge components, and standard 3 ton 4 × 4 lorries formed the main Bailey bridge platoon equipment. The standard platoon carried enough material to construct a Class 40 bridge, 130 ft long. If the heavier Class 70 version was required, then additional parts were needed. These were carried by the Heavy Bailey increment section of four 3 ton lorries. Pontoons for floating Baileys were transported on the standard pontoon lorry, but with the advent of the 50/60 pontoon, a special 5 ton four-wheel transporting and launching trailer was developed and put into production in early 1944. A typical tractor for this trailer was the FWD SU COE or the heavier White 666 or Mack NM 6; four sections of five tractor/trailer combinations forming a platoon.

In addition to the special bridging vehicles, standard mobile cranes were provided to lift heavy equipment. These gradually replaced the

*Pontoon equipment, drawn by Autocar U-8144T tractors, waits to move up to the Rhine at Remagen.*

pre-war 3 ton derrick. The derrick was much cheaper to purchase and operate than a mobile crane as it consisted of a simple welded framework attached to the rear of a standard 6 × 4 chassis, it was hinged, allowing it to fold forward to reduce travelling height. Lifting power was provided by the standard type of winch for this class of vehicle. The major disadvantage was that there was no slewing facility. The derrick was usually fitted to the Crossley IGL8, Guy FBAX or Leyland Retriever chassis. Motor-boat tugs formed part of the bridging company complement and were used for manoeuvring pontoons into position. A special 2 ton four-wheel transporting and launching trailer for this tug was towed by a standard 3 ton lorry.

The Bailey bridge revolutionised military bridging, and over half a million tons of bridging was produced during World War 2, large quantities were supplied to the Canadian and American armies in Europe and it saw world wide service with the British. The amount and type of bridging equipment varied from theatre of operations to theatre of operations. Obviously, very little was required in the Western Desert but vast quantities were needed in Europe. Bridges had been destroyed by the Allied Air Force, by the Resistance movements and, what was left, by the retreating Germans, and they had to be replaced. In Italy, the swift rivers often made rafting and pontoon bridges impracticable. Thus, the make up of a British bridging company was kept flexible so it could be quickly reorganised to suit changes in tactical requirements. The normal establishment was an HQ detachment including Royal Engineers, a workshop platoon and ten task platoons. In NW Europe these normally consisted of the following platoons, three Bailey bridge, two Bailey pontoon, two FBE Mk III, one close support raft, one assault and one heavy Bailey bridge section. The whole consisted of 450 vehicles and 750 men commanded by a Major.

Some idea of the amount of equipment needed in NW Europe can be gained from the fact that over 800 extra 3 tonners were transferred to assist in moving bridging material in preparation for the clearing of Antwerp and the Scheldt Estuary.

The standard American bridging system was the 'Treadway'. A self contained bridge section was carried on the Brockway 6 ton 6 × 6 bridge construction truck. Each vehicle had an integral hydraulically operated swinging gantry hoist, to handle the bridge components mounted at the rear of the chassis. Pontoons (American 'pontons') came in single units, 35 ft long by 8 ft wide, carried singly and upside down on a 25 ton ponton semi-trailer, drawn by the Autocar U-8144T 5 ton 4 × 4 truck tractor. The Autocar U-8144T was also supplied to the British Army. These pontons were very unwieldy to unload and launch. Large inflatable rubber boats, carried in normal trucks, were available to double as pontons.

Prior to the introduction of the Brockway, the FWD YU COE, fitted with a simple derrick and winch, was used to handle Treadway components and there was a wide range of mobile cranes, including crawler mounted types, available to the Corps of Engineers.

Apart from specialist bridging equipment, the engineers of all armies had a wide variety of equipment for earthmoving, road building and general construction work, these were invariably adapted from their civilian counterparts.

# MISCELLANEOUS SPECIALIST VEHICLES

In addition to the various types already described, there was a wide range of other specialist vehicles, whose presence was no less vital to the efficiency of a fighting force, but, because of the nature of their task, their numbers were a small percentage of the total.

The method of locating enemy artillery positions had long been for observers, at accurately surveyed locations to plot the flash and/or the sound of the gunfire. This information was fed back to a central information post, where the bearings taken by the various observers could be plotted, and enemy gun positions pinpointed; the map references were then relayed to supporting artillery to bring down counter-battery fire. The British Army had a special house-type body fitted to the Leyland Retriever chassis for sound ranging, recording and plotting. The Germans used similar equipment, installed in special Kfz 62 house-type bodies fitted to pre-war standard light 6 × 4 chassis or to the later 6 × 6 diesel engined IE LKW; while their forward survey and spotting parties used the Kfz 63 and 64, which were not unlike the pre-war British Morris 6 × 4 reconnaissance car and carried a crew of five or six, dependent on their role, plus specialised equipment in the rear stowage lockers. The British, however, made do with standard GS vehicles.

One of the unpleasant features of military life in the field was the appearance of various sorts of lice. In order to rid body and clothing of these visitors, special decontamination vehicles and mobile showers or bath houses were attached to

*The famous 'Queen Mary' RAF aircraft transporter was towed by a variety of prime movers, in this case an early Crossley 'Q' type known as a 'Donkey'.*

every division. Decontamination of clothing was carried out by fumigation and steam cleaning. In British service, this equipment consisted of a steam boiler and drum mounted on a base frame that could be carried in any standard 3 ton GS lorry, or demounted for static use. Shower units were normally trailer mounted and towed by 3 tonners. The Germans, however, had special van bodies, for both clothing (Kfz 63) and personal decontamination (Kfz 62b), fitted to the medium 6 × 4 Henschel 33Gl. The American equipment was, in general, carried in 6 ton semi-trailers towed by Autocar 7144T 4 × 4, 4–5 ton tractors. These semi-trailers had fold down sides and could be linked together to form larger capacity units.

Water supply has always been a prime requirement for any army, and during World War 2 a regular drinkable supply had to be provided, regardless of climatic or geographic conditions. Once a source of water had been tested and its integrity assured then distribution was undertaken by bulk tankers. These tankers were equipped with power and hand pumps to draw up or pump out water and with filters and sterilisation equipment, it was also usual to have a row of taps to enable troops to fill up water bottles. The standard American tank was the 700 US gallon capacity unit, on the GMC $2\frac{1}{2}$ ton 6 × 6 chassis, and a 1,000 US gallon tanker was also

Top right *Typical specialist workshop body fitted to a closed cab GMC 353, which was superseded in July 1943 by the open version. No winch fitted.*

Bottom Right *The Opel 'Blitz' was produced in both 4 × 2 and 4 × 4 versions. Over 70,000 chassis were supplied to the Wehrmacht between 1937 and 1944. Many specialist bodies were fitted and this photograph is of a 'house-type' medical body.*

USA 0061708

WH - 31874

*The Royal Engineers used air-compressor versions of the 15 cwt for a variety of tasks. This pre-war Morris CS8 is fitted with an Air Pumps Ltd compressor to power a variety of pneumatic tools including shovels, drills and hammers.*

available on the Diamond 'T' 4 ton 6 × 6 chassis, together with a 1,500 US gallon capacity semi-trailer. There was no American equivalent to the British 15 cwt 200 gallon tanker, but units were issued with 250 US gallon water trailers.

The British had a variety of 3 ton 4 × 2 chassis on which was mounted a 450 Imp gallon tank unit, these were used for distribution work, typical chassis being the Canadian Dodge T 110-L, the CMP 3 ton 4 × 4 and the ubiquitous Bedford OY. Battalions were issued with 200 gallon tankers, commonly known as Unit Water Carts, fitted to the 15 cwt 4 × 2 Bedford MW, the Morris CS8, the Canadian Dodge D15 and the CMP C15A 4 × 4. There were minor design changes made during the production of the British 200 gallon tanker. Initially, the power-driven pump was attached to the front bumper and meshed into the crankshaft starter-handle dogs when required. In the later models, a PTO from the gearbox drove the pump. The shape of the cross-section of the tank also changed, from oval to slab-sided, and the capacity was increased to 230 Imp gallons. The canvas tilt and tubular support, intended to disguise the water tankers as GS trucks and provide some protection from the sun, was often removed in service. The British also had 90 gallon two-wheeler water trailers based on a pre-war design.

Another small tanker was the Bulk Contamination Vehicle – built in limited numbers when the threat of German invasion of the British Isles seemed imminent. Consisting of a 100 gallon cylindrical tank mounted on the Morris CS8 chassis, it was intended to pump pollutant into water mains to deny drinking water to the invader. Whether this would have been carried out is a matter of conjecture as, undoubtedly, the native population would also have suffered.

Fuel was the other vital fluid needed to sustain a mechanised fighting force, and tankers had to be provided to distribute it from dumps and railheads. In the period August 28 to September 12, 1944, the British Guards Armoured Division advanced 495 miles, this consumed 892,000 gallons of fuel, over a 1,000 lorry loads, and in one particular night 130 lorry loads of fuel were issued.

It was policy to issue front line units with fuel in cans transported by normal cargo trucks, otherwise a large number of tankers would have been required. The bulk fuel tankers collected fuel from pipeline or rail terminals and delivered it to can filling points. Most nations utilised their medium truck chassis to carry fuel tanks of 800 to 1,000 gallon capacity. The standard British tanker was the Bedford OY with an 800 gallon Thompson built tank unit, equipped with filters,

*A late production Guy Ant with enclosed 'Vixant' type cab and light warning body – a radar vehicle used with AA Regts.*

*An MW 200 g water tanker of the Aldershot District filling from a local pond. Samples are lined up on the bonnet to undergo purity checks.*

pump and delivery hoses. This was supplemented by over 400 Bedford OXC-Scammell 1,750 gallon semi-trailer units. The Canadians also produced an 800 gallon tanker, initially on the CMP 3 ton 4 × 2 Ford F602L, and later, in larger numbers, on the Chevrolet C60L 4 × 4. The American standard bulk fuel tanker consisted of two 375 US gallon demountable units on either the GMC or Studebaker $2\frac{1}{2}$ ton 6 × 6 chassis. These were also supplied to the British.

In Germany, the Henschel 33G1 6 × 4 medium truck with a 3,600 litre tank saw service as the Wehrmacht's main bulk fuel carrier, backed by the smaller capacity Kfz 385 on the Opel Blitz 4 × 2 chassis.

With the great strides made in aerial photography during the early war years, and the introduction of tactical fighter reconnaissance support for the ground forces, speedy processing of film had to be made available in the field, so that commanders could have up to the minute information on the disposition of opposing forces. Facilities had to be provided to handle, develop and process film, and to convert this information into maps as required.

The early British equipment was installed in 15 ft house-type bodies on 6 × 4 3 ton chassis, but as the equipment was bulky and heavy, a 21 ft body was designed to suit the Foden DG/6/12 6 × 4 10 ton chassis used on Foden's pre-war diesel-engined commercial lorry. The Leyland Hippo Mk 2 could also carry this series of bodies.

There were four vehicles forming a Field Survey Reproduction Section.

The Auto Processing vehicle received the film from the reconnaissance aircraft cameras, usually in 200 ft rolls, and developed, fixed and dried the exposed film. A power operated auto processing machine was installed with developing, washing and hypo tanks, together with a drying drum. The usual photographic dark-room lighting and accessory equipment was also fitted. Power for all these vehicles was supplied from individual 24 Kw generator trailers.

The rolls of processed film would then be transferred to the Enlarging and Rectifying lorry. Also a 21 ft house-type, this contained special enlarging and rectifying apparatus, and here negatives could be enlarged to a suitable size for later reproduction as maps. Rectification of aerial photographs was often necessary to eliminate distortion of the vertical overlap film, caused by tilting of the camera, due to slight movement of the aircraft. Oblique aerial photographs were not used to make maps, but constituted a form of intelligence gathering.

The next stage was to produce printing-plates from photographic negatives. This process was carried out in the Photo Mechanical lorry, which differed from the previous vehicles in that the

*The Canadian-built Dodge D15, 15 cwt, 4 × 2, 200 gallon water tanker was also fitted with a CMP 15 cwt GS body and used by the British Army in large numbers.*

*Special bodies were fitted to many 3 tonners and this bread delivery van on an Albion KL 127 chassis delivered loaves from field bakeries.*

*Small numbers of horse ambulances were required despite complete mechanisation of operational units, and bodies of pre-war design were fitted on Bedford OY chassis.*

sides could be opened out from 8½ to 15 ft to give greater working area. It was equipped with a water supply, sinks, developing tables, whirler for washing and drying plates, a plate proof-press and demy printing down frames as well as arc lamps.

Once the zinc map printing-plates had been proofed and found satisfactory they would be passed to the Printing lorry for quantity production up to demy size, i.e. 22½ × 17½ in.

The width of the printing body could also be extended to 15 ft and, because of the printing presses installed, it was fitted with levelling and steadying jacks at each corner. The press was a Crabtree rotary self-feeding offset machine, capable of producing 4,000 copies per hour, extra colours requiring another run per colour. It was also fitted with work tables, racks for printing-plates and storage lockers.

Powered by a Gardner 6 cylinder diesel, developing 102 bhp at 1,700 rpm, weighing a maximum of 15 ton 7 cwt, the DG/6/12 was capable of a top speed of 30 mph.

Despite the provision of these field reproduction facilities, the British Armoured Divisions literally ran off the map in the dash for Brussels and Antwerp and in exceptional circumstances had to rely on maps obtained locally, including Michelin guides! These vehicles were also used by the Canadian Army in Europe and the air forces had similar mobile facilities for their tactical squadrons.

The American Corps of Engineers used a basic house-type body on the GMC 1½ ton 6 × 6 chassis in a number of roles for photographic and map reproduction, but due to the smaller body capacity, seven vehicles were required to fulfil the same task as four Fodens. These were laboratory section, map layout section, photographic section, plate grainer section, plate process section, camera section and press section. Semi-trailer vans were also available, with the 2½ tonners housing camera and press equipment and 10 tonners equipped with the heavier printing equipment, etc.

Like the Allies, the Germans equipped standard house-type bodies for photographic reproduction and printing work, utilising the light 6 × 4 Kfz 62 and the medium Krupp L3H 6 × 4 Kfz 354 as well as variants of the closed Opel Blitz Kfz 305/90 and 91.

There was a requirement for mobile office and printing facilities in the field, as no army can function without some form of administration despite opinions expressed by the front line troops.

At the beginning of hostilities, there were a number of special office trucks available to both sides – chiefly the house-type, fitted out with desks, chairs, filing cabinets, typewriters, dup-

*A standard Bedford QLC with mobile kitchen equipment installed. These vehicles provided cooking facilities for mobile formations usually one to each squadron.*

*The Austin K5 shown here is fitted with a Westex Recorder body used by heavy A/A regiments in conjunction with locating radar.*

licating machines, etc. Pent houses could be rigged between vehicles to provide extra covered workspace, and mobile generators provided power for lighting and heating. However, it was soon realised that special office bodies took more time to produce than their operational necessity warranted, and standard GS or cargo trucks replaced them as the war progressed. The modifications were kept to a minimum, usually consisting of provision for lighting and translucent panels let into the canvas tilt to act as windows, with side pent house extensions.

Early British house-type office bodies were either versions on the Bedford OY and QL or the CMP Lindsay bodies (OFF-1 and OFF-2) on the Chevrolet C60L 4 × 4 chassis.

A unique feature of the German Army was the widespread use of motor-buses for operational purposes. Apart from the ambulance role already mentioned, the Wehrmacht-Omnibussen were used as command and office vehicles, their roomy interior making them eminently suited for this role despite their lack of manoeuvrability.

The British Army, did, in fact, use buses in the crisis days after Dunkirk. They were used for transporting troops, as command vehicles and as lines-of-communication vehicles (LCVs), but the buses were all impressed from civilian operators, and it was suspected that many of the best vehicles remained in their owners' hands, despite inspection by Government officials.

A special office body was fitted to the Morris CS8 15 cwt 4 × 2, intended for issue to infantry battalions and brigade HQs, these were gradually replaced by 'Vans 15 cwt', this was a hard top version of the standard 15 cwt GS and the body tilt was a permanent fixture.

Apart from administrative or office trucks, command vehicles were required for unit commanders from corps down to regimental or battalion level. In the armoured formations of the British Army, these were wheeled armoured vehicles or ACVs. The ACV originated with experiments carried out by the 1st Armoured Division on armoured 15 cwt GS trucks before the war. Guy Motors produced the first proper ACV on their 3 ton 4 × 4 Lizard chassis, but only 14 were produced before they were superseded by a larger vehicle on the Matador chassis known as the Dorchester. The Dorchester was a popular vehicle, not only with the British, but also with Rommel who made good use of captured examples in the Western Desert. An improved version on an AEC 6 × 6 was in production by the end of 1944; and first unit to receive them was the 7th Armoured Division after the Rhine crossing.

Regimental and battalion commanders used White scout cars and half-tracks specially fitted with mapboards, radios, etc. These vehicles suffered from lack of overhead protection as did many of the German semi-tracks. Nevertheless, the White scout car was popular as a reconnaissance vehicle with the British supporting arms because it was reasonably roomy. The REME particularly favoured it for battlefield recovery reconnaissance of damaged vehicles. However, it went out of production in October 1943. The British War Department decided a replacement was needed, and came up with a prototype design, the CMP field artillery tractor, with a fully enclosed armoured body, which was known as the 'Cupola'. The Canadians carried out some drastic re-design and came up with a 15 cwt 4 × 4 armoured truck based on the Otter LRC chassis. Apart from those ordered for the Canadian Army, contracts for 3,000 were placed by the British late in 1943. These vehicles saw service in NW Europe into the post-war years. A total of eight fully-equipped men were carried, and floor plates could be lifted to reveal foot wells. Folding seats were provided which, when not required, were stowed in the wells leaving the floor clear for cargo. Alternatively, a light AA gun could be mounted or two stretchers fitted to give an ambulance role. Front armour was 14 mm ($\frac{9}{16}$ in.) with 6 mm ($\frac{2}{8}$ in.) sides and rear, however, although the cab had an armoured roof, the body had relatively low sides and no top protection.

An armoured lorry was developed from the Deacon SP 6 pdr anti-tank gun which, in turn,

*An American Army mobile kitchen on a $1\frac{1}{2}$ ton 4 × 4 – the troop seats have been folded outboard to provide extra working tops.*

was built on the Matador chassis. The new lorry retained the armoured cab for the driver, with a lightly armoured cargo body in place of the 6 pdr turret. This truck was intended for use as an ammunition forward supply vehicle, for the self-propelled artillery regiments of armoured divisions.

Mobile living/office quarters were often provided for senior field officers above the rank of brigadier. Those of the Italian Army were most luxurious and consequently highly prized items of booty. Probably the most famous of these was the one captured by the British in the Western Desert and refurbished for use by General Montgomery. The body was removed from its Italian chassis and refitted to a Mack NR-4 10 ton 6 × 4 chassis. The caravan was divided into separate bathroom and living quarters. A second captured Italian caravan, mounted on a Leyland Retriever 6 × 4 3 ton chassis, was used by General Montgomery to complement the larger NR-4.

*This example of an FWD – SU-COE fitted with a British medium artillery tractor body has been extensively modified to take snow-plough equipment.*

Both caravans served through the North African and Italian campaigns and were shipped to Britain for refurbishing prior to D-Day. They continued to be used by, the now, Field Marshal Montgomery until VE Day, and are now on display in the Imperial War Museum, London, together with the Field Marshal's mobile operations room.

A handful of luxurious caravans were produced by the REME workshops at Abbassia early in 1944, and fitted to CMP C60L 4 × 4 3 ton chassis. The first type was panelled in Indian red mahogany with convertible bed settee, ice boxes, fitted wardrobes etc., and had canvas side penthouses. The second type was less luxurious and more functional. Construction was of T & G boarding, and the forward compartment was fitted out as a bathroom and lavatory with living/office accommodation to the rear, no penthouses were carried. Three of each type were produced, and were subjected to criticism, as it was thought unnecessary to mount such bodies on 4 × 4 chassis when front line units desperately needed the chassis for operational purposes; it was suggested that, as the war was being waged in a more civilised environment, there should be plenty of vacant chateaux in which senior officers could be billeted. In any case, a four-wheeled trailer caravan was considered to be a more suitable vehicle for this purpose as senior officers generally used staff command cars to travel.

A multitude of staff cars have been produced since the introduction of the self-propelled vehicle into military service, the majority of which were basically civilian models and not intended for use in the field. Many staff cars have become well known, and from World War 1 we have the Prince Henry Vauxhall tourers, the Rolls Royce Silver Ghosts, the RFC Crossleys and the Ford Model T in its many guises. In World War 2, the Humber Snipe saloons and tourers were a familiar sight, General Montgomery's 'Old Faithful' probably being the best known of all. The militarised Ford Model 62

*The sound-ranging and plotter body was used in conjunction with sound-locating equipment to determine the position of enemy guns. A comparatively rare vehicle. The Retriever now sports a windscreen as a sop to driver comfort.*

*This Austin K6 was one of a small number fitted with caravan bodies for senior commanders by Carbodies Ltd for the RAF Tactical Air Forces in NW Europe.*

*Another militarised civilian lorry was the Foden DG/6/12 10 ton 6 × 4 fitted here with a printing body for map reproduction, the extensions fold away when in transit.*

with 30 hp V-8 and 'squared' front end, cut away wings and large section tyres became the W0A1/A and was used in large numbers. Many saw service in the desert campaigns and were equipped with extra water and petrol can stowages and fitted with desert tyres. The British and Canadian heavy utility class was equal to the American command cars or 'carry alls' but, with the exception of the Humber 'Box', were 4 × 2 drive. These cars had the advantage of being roomier than saloons and could accommodate mapboards and other staff impedimenta, some versions carried radio sets. A 'scalped' version of the Canadian Ford C11 ADF heavy utility, used by Field Marshal Alexander in North Africa and Italy, is now preserved in the Canadian Forces War Museum in Ottawa.

Despite their vast motor car industry, the Americans standardised only five types of 'light sedans' and two types of medium, settling instead for ½ ton and ¾ ton 4 × 4 command and reconnaissance vehicles produced by Dodge and still a familiar sight in some armies and in the hands of preservationists. However, these vehicles lacked space and adequate weather protection, so the carry all, itself a panel truck or van with windows, was equipped with the usual staff furniture for use as a command car. Small numbers saw service in the British Army, chiefly in Italy and the Far East.

The Germans, on the other hand, and despite the Schell rationalisation programme of the 1930s, had a bewildering array of light, medium and heavy 4 × 2 and 4 × 4 cars. Many were used in combat as opposed to staff or command roles and incorporated several interesting and unique design features.

The Steyr 1500A/01 was a command car in the heavy 4 × 4 Kfz 21 class. It was based on the 1½ ton 4 × 4 light truck chassis suitably lightened. The front-wheel-drive differential was incorporated in engine sump, and was powered by an air-cooled V-8 3½ litre engine developing 85 bhp. Front wheel suspension was by torsion bars with half-elliptic leaf-springs at the rear. This vehicle was highly regarded by the Germans, and was produced in quantity by Steyr and sub-contractors such as Auto Union.

An example captured by the Royal Sussex Regiment at Cape Bon in Tunisia, was shipped to the WVEE for test. The car had been used by General von Arnim prior to its capture. On test it was found that the suspension, stability, acceleration and hill-climbing capabilities were better than its British equivalent, while fuel consumption at 30 mph was 14.9 mpg – a marked improvement. However, the steering was considered to be over heavy and the oiltight, dust-excluding concertina-type synthetic rubber boots enclosing the ball joints and drive couplings were thought to be very vulnerable to damage during cross-country work, and a preference was expressed for felt or leather seals – times and design thinking have changed since those days.

The heavy class Auto Union/Horsch 1a was introduced as a standard design intended to replace a variety of 6 × 4 and 4 × 2 vehicles, however, it was expensive to produce, complicated to maintain and did not entirely replace much simpler vehicles already in service. It had independent wishbone suspension on all wheels, four-wheel drive, and steering on all wheels, although the rear-wheel steering was only intended for low speed work and could be disconnected as required. The 3,823cc V-8 engine was also unorthodox, in that the valves lay horizontally and the cylinder head joint was in two planes at 130° to each other – this meant a special sequence of tightening down the head bolts to ensure a gas-tight seal. The valve operating gear was a veritable 'rats nest' of 16 overlapping springs and rocker arms lying in between the two cylinder heads. The gearbox was also unusual, in that lubrication was by pump, and forward emergency low gear was obtained by 'reversing reverse', the gearbox was also criticised as being heavy to operate. Apart from use as a command/staff car, there were other variants, such as ambulance, gun tractor, light AA etc.

Before the advent of radar, location of enemy aircraft depended on sound location and, at night, searchlight illumination of the hostile target for the anti-aircraft gunners. Special trucks were designed in the 1920s and 1930s to transport and power mobile searchlights. In some cases, elaborate lifting gear was provided together with extensive stowage space and crew accommodation. A typical example being an early version of the Guy FBAX, power being supplied by a 24 Kw generator, belt driven from the gearbox. Dictates of economy soon resulted in simpler designs, and the British searchlight lorries were really basic GS trucks with relatively simple role modifications. The standard searchlight was the 90 cm projector, mounted on four minuscule tracked bogies, and trained on to the target by a handwheel on long outrigger arms. This equipment could easily be accommodated in a conventional 3 ton 4 × 2 lorry, and the three standard vehicles in this class were basically civilian in concept. These were the Thornycroft ZS/TC4, which had a generator mounted in front of the engine, driven from the crankshaft extension. The projector was loaded into the body by muscle power, up simple channel ramps which, when not in use, were stowed under the body between the chassis fillers. Seats for the

*The American Brockway 10 ton 4 × 2 tipper saw service with the British Army in relatively small numbers.*

*Another rare bird in British service was the Mack $17\frac{1}{2}$ ton 6 × 4 dump truck.*

*Many specialist bodies were fitted to CMP chassis but were eventually phased out in favour of standard GS bodies, role equipped. The Chevrolet C60L carried a Lindsay house-type office body.*

crew of eight were bolted to the body floor. The Guy PE and Tilling Stevens TS 19 and 20 models were similar in equipment, but being basically petrol-electric no extra generating equipment had to be fitted.

Several of the standard WD 6 × 4 3 ton chassis were fitted with searchlight bodies, these being the Guy FBAX, Leyland Retriever, and Crossley IGL8, as well as the Fordson Sussex. The 24 Kw generator was driven by belt and protruded into the flat floor body under the transverse crew seats. Loading was by manhandling up ramps. Although these vehicles served throughout the war, a larger 150 cm projector was introduced into service and was trailer-mounted, being too big for the 3 tonners and requiring more power than could be provided by a 24 Kw generator. By this time both projectors had also acquired various radar antennae which made transportation less easy.

The American standard searchlight vehicle was the GMC AFWX-354 COE $2\frac{1}{2}$ ton 6 × 4, powered by the Model 256 engine in the US Army but supplied for French contracts with a smaller capacity Model 248 engine. Many of these ended up in Britain after the fall of France. The crew were accommodated in an extension to the driver's cab, and power was supplied by a towed generator trailer. This truck could also be used in a secondary cargo role.

Finally, before leaving the specialist vehicles

and taking stock of load carrying trucks, it seems a logical step to look at tipper or dump trucks.

The tipper or dump truck performs an unspectacular but essential task as part of the engineering branch of any army. Their main task was to transport road making materials, rubble crushed stone, etc., but quite often they were pressed into service as general transport vehicles. In the British Army, the Tipper Company was part of the RASC organisation and consisted of an HQ and four platoons. Each platoon, commanded by a lieutenant, was further sub-divided into three sections of nine tippers, each with a corporal in command of the section. Tipper companies rarely operated as a complete unit, but were allocated piecemeal to engineer units throughout a division as required. It was not unusual for single vehicles to be detached from their section for long periods, and at one stage of the NW European campaign, the platoons of one company were up to 200 miles apart. As already stated the main role of the tippers was to transport road making material, and this became a continual task, particularly in the flooded Low Countries where the long lines of communication back to Normandy put an enormous strain on the often improvised road network. In the Guards Armoured Division, during the advance, a tipper loaded with rubble was interposed between each tank. When a crater was encountered, the tippers filled it in and the advance continued, the tippers being replenished from demolished houses etc. Typically, in a period of 24 hours, one tipper section moved 500 tons of rubble in 171 loads. During the hard winter of 1944/45 tippers were engaged on grit spreading duties on the icy roads.

The majority of British tippers were militarised civil vehicles in the 3 ton class, usually end tippers but a number of three-way tippers were also produced; the hydraulic tipping gear being the product of specialist firms such as Edbro.

Tipper bodies were deliberately low sided to prevent over-loading. Typical examples were the Dennis and Thornycroft WZ/TC4, the latter being one of the three-way variety. A tipper version of the Austin K-6 was also produced. The Bedford QL(APT) tipper was the only British 4 × 4 in this class. It was designed to double as a load carrier continuing to equip tipper companies well into the post-war years. The Fordson Thomas 7V, although dimensionally similar to the 3 tonners, had a 4 cu. yard steel body and was rated at 6 tons, being used chiefly by the RAF for airfield constructional work. Several tippers were available from Canada, the Dodge T110L-6 4 × 2 3 ton dump was a modified conventional type produced in quantity with an all-steel body. It was followed by CMP variants on the Ford F60S and Chevrolet C60S chassis, with Garwood teledraulic hoist equipment and Budd all-steel 4C1 and 4F1 dump bodies.

The Americans had a wide range of dump trucks available, either based on civilian designs or standardised military vehicles. The smallest civilian design was the Chevrolet 1½ ton 4 × 2 dump, followed by the International 2½ ton 4 × 2 and 5 ton 4 × 2 models. The military dump bodies were similar to cargo bodies in that, with the exception of the Diamond 'T' Model 972 4 ton 6 × 6, they had stake sides incorporating folding troop seats. Both Chevrolet 1½ ton 4 × 4 and GMC 2½ ton 6 × 6 dump trucks were supplied to the British under lend-lease, as was the enormous Mack 17½ ton dump truck based on the NO2 chassis. Many of these military tippers/dumpers found their way into civil engineering companies after the war and soldiered on until they could be replaced by new civilian models.

# LOAD CARRYING VEHICLES

L4555904
5
7
907 JJO

Generally speaking most military load carrying vehicles could be divided into three main classes, light – up to 1 ton capacity, medium – $1\frac{1}{2}$ to 4 tons, and heavy – above 5 tons. These were further sub-divided into those capable of cross-country operation and those, the remainder, that had limited off-the-road performance. The description of the load carrying role varied. The British used the term General Service (GS) as a suffix to the vehicle description, e.g. 'Truck 15 cwt 4 × 2 GS' (Bedford MW) or 'Lorry 3 ton 4 × 4 GS' (Austin K5). Truck being used for vehicles up to 15 cwt payload and lorry for the 3 tonners upwards. The Americans used the term Truck for all load carriers followed by Cargo, thus 'Truck, Cargo, LWB, Cargo, $2\frac{1}{2}$ TON 6 × 6' would be qualified by the manufacturer's name and model, the LWB (Long Wheel Base) being used if there were two basic chassis lengths. In Germany a heavy all-wheel drive cargo truck would be described as Schwere geländegänziger Lastkraftwagen, often abbreviated to s.Lkw (A-Type).

In Britain, the lightest GS vehicles were classed as, 'Trucks 5 cwt 4 × 2 Light Utility', commonly known as 'Tillies', and were produced by Austin, Hillman, Morris and Standard, based on their individual 10 or 12 hp pre-war passenger saloons. These vehicles were essentially a compromise, lacking a reasonable cross-country performance and largely replaced by the Jeep in front-line units. Their light constructions led to distortion of axles, chassis and suspension, and the low power/weight ratio led to excessive use of the gears, inevitably causing gearbox failure and causing the king-pins and bushes to wear out.

The 8 cwt GS class had been dropped from British production in an attempt to rationalise production plans. Despite this there were still relatively large numbers in service, although many were lost at Dunkirk. These vehicles were a smaller edition of the 15 cwt and had better weather protection for the driver, full width fixed screens, half-doors and side curtains. Morris-Commercial produced the PU/MK2 4 × 2, and the PU8/4 4 × 4 in smaller quantities, while Humber had a 4 × 2 based on their Snipe chassis, and a standard 8 cwt body mounted on the 4 × 4 chassis also used for the 'Box' heavy utility and light ambulance. The Ford WOC-1 was a modified US Ford design utilising the chassis and front end of the 1940 US V-8 saloon, this vehicle was put into limited production only. The Canadians produced a parallel range on the CMP 4 × 2 C8 and F8 chassis as well as a 4 × 4 version on the Dodge T212 D8A.

The 8 cwt body was produced either as a personnel carrier with tip-up seats for three or 'Fitted For Wireless', in which case, accommodation for only two personnel was provided and the entire tilt could be removed for use as a tent with extended ground spikes. The 4 × 4 version had a better cross-country performance but its limited carrying capacity resulted in the 8 cwt being phased out of production in favour of the 15 cwt, which cost little more in time or money to produce.

The 15 cwt GS truck was the first type of vehicle to go into mass production when full mechanisation of the British Army was undertaken in the 1930s. It was intended to carry the heavy personal equipment of an infantry platoon, such as blanket rolls, large packs, extra ammunition and rations. No 'cape-cart' tilt was fitted to the pre-war vehicles, and only the driver had a protective hood with rudimentary 'aero-screens' and canvas side curtains. The original standard 15 cwt GS body sides were hinged just above the wheel arch, this allowed the upper half to fold down to give a good field of fire for a proposed machine-gun mounting. This was eliminated on later production runs. Apart from the GS body, the specification called for the fitment of a WD pattern drawbar, low-pressure large-section cross-country tyres, stowage boxes for tools, overall chains for the rear wheels, and POW (Petrol Oil Water) can carriers. The early design could take either one 2 gallon can or two 1 gallon cans. With the advent of the Jerrican, the carriers were modified to accept this universal container. The mechanical components were basically

Previous Pages *Bedford OYD 3 ton 4 × 2 GS. Most numerous of the British Army 3 ton 4 × 2 class, also fitted with many types of specialist body. This preserved example has to comply with current lighting regulations and blackout hoods are not fitted (see p. 134 for engine compartment).*

*The Bedford QLC 6 ton 4 × 4 – 2 with permanently-coupled semi-trailer doubled the payload of the normal QL truck. A ton of freight could be carried in the small body behind the tractor cab.*

*The American equivalent of the Bedford was the Studebaker 6 × 4–2 4/5 ton cargo truck supplied under lend-lease to the British Army, and to the RAF with different pattern of semi-trailer.*

commercial and usually common throughout a manufacturer's range of 15 cwt to 3 ton trucks.

The low silhouette and short length meant that most of this class were semi forward-control, and the gearbox cowling usually humped up between the driver and his mate. Seats were bolted directly to the floor, and, with legs stretching out to the pedals, the driving position was not unlike pre-war sports cars such as the MG Midget.

Spare wheel carriers were not fitted to vehicles equipped with 'runflat' tyres.

Morris-Commercial were first into production, in 1934, with their CS8 Mk 1 which was in quantity service by 1936. Minor changes were made to the shape of the front wings, bonnet and radiator during its production life before it was replaced by the C4 Mk 1 which had better weather protection. Weather protection introduced on all 15 cwts in 1943 included metal half-doors and full width windscreens. The C4 was also fitted with a 4 cylinder engine, replacing the 6 cylinder of the CS8, but giving the same 60 bhp. In November 1943 the C4 Mk II replaced the Mk I in production, the only difference being that the back axle had been moved 9 in. rearwards, increasing the wheelbase to 8 ft 11 in., to obtain better weight distribution. Other dimensions were unchanged.

By October 1944, the Morris 4 × 2 15 cwt was replaced in production by a 4 × 4 – the C8/GS, based on the company's Quad chassis but without the winch. Although the engine gave another 10 hp and a five speed gearbox was fitted, a fully laden C8/GS was $\frac{3}{4}$ ton heavier than its 4 × 2 predecessors and always seemed sluggish and very noisy. It could be used in the FFW role, be fitted with a compressor or with a light warning radar house body.

Bedford had been successful in getting their 15 cwt adopted for production in 1937, following trials with prototypes in the previous two years, and deliveries of the MW started in August 1939. The slab fronted MW was the most numerous of the British built 15 cwts, and served into the post-war years, uprated to a 1 tonner. A chassis with fully sectioned engine, gearbox, and rear axle now driven by electric motor, still serves as an instructional aid in the Southampton TA Centre. Apart from the standard GS body (MWD), this chassis was used as an anti-tank portee, water tanker (MWC), 20 mm Polsten A/A mounting (MWG), FFW (MWR), 2 pdr anti-tank gun tractor (MWT) and as a van (MWV). The GS body could also carry the KL machinery (light) welding apparatus or an air compressor and tools.

Another early entrant into 15 cwt production was Guy Motors who introduced the Ant in

USA
USA

Top Left *The Dodge weapons carrier $\frac{3}{4}$ ton 4 × 4 shown here with winch fitted. It was a most useful infantry-platoon vehicle and many were supplied under lend-lease. Still in service in various armies of the world, it is a popular vehicle with restorers.*

Bottom Left *The command-reconnaissance WC 56 version of the Dodge T214 also saw service in considerable numbers. Provision was made for a radio installation and a winch was fitted to the WC57 variant.*

Top Right *The Humber FWD 4 × 4 heavy utility, commonly known as the 'Box', it was the only British-built type in this class and had a very good cross-country performance. The chassis was used for a light ambulance, 8 cwt PU and light recce car.*

1935. A full civilian-type cab was fitted when the chassis mounted the light warning body. The Quad Ant 4 × 4 15 cwt started to come off the production lines in January 1944 having replaced the Quad Ant artillery tractor and using many of its components, but numbers in service were never very large.

The Ford Motor Company also produced an indigenous 15 cwt. The WOT-2 series commenced in 1940 and ranged from the 2A to 2H, covering changes in windscreen, bonnet and body style. It would appear that the WOT-2 was produced in batches, as and when material and parts were available, suggesting that the company's other products had priority. A small number of 15 cwt vans were also produced between October 1940 and July 1941 on imported US EO1 chassis.

The Austin Motor Company and Commer had produced 15 cwts before the War and these were still in service at the outbreak of hostilities. The Austin BYD was based on the Ascot taxi chassis with artillery-type spoked wheels and the Commer Beetle was similar in appearance to the Bedford MW but replaced by the Q2 and Q15 models used mainly by the RAF.

As well as the Dodge D15, the Canadian

*A Bantam 40 BRC supplied to Britain and shown with extra stowage and Lakemen/Bren mounting, in service with the 6th Armoured Division.*

industry provided large numbers of 15 cwt vehicles in their CMP range. These were more luxuriously appointed than their British counterparts, having full metal cabs, and doors with sidescreens. Earlier, quantities of Chevrolet 1311 × 3 15 cwt 4 × 2 trucks had been delivered to the Egyptian Army and to the British and Indian forces in North Africa. Several of these vehicles were acquired by the Long Range Desert Group and, after stripping down to the bare essentials, were used by several LRDG patrols, until replaced in 1941 by the larger Chevrolet 1533 × 2 30 cwt 4 × 2.

The Canadian 15 cwts, due to their greater power-to-weight ratio, were generally more reliable than their British equivalents, although for a period radiators gave trouble because of their somewhat flimsy construction.

The 4 × 2 version of the 30 cwt load carrier was virtually a militarised version of the pre-war civilian light lorry and as such was produced in small quantities by most British companies in the early War years. Most of these vehicles retained the civilian-type closed cab, an exception being the Austin K30. The only British 4 × 4 30 cwt was the Ford WOT-8 which had many parts in common with the WOT-6 3 tonner. The Bedford OXD was probably the most numerous in this class, as the chassis was also used as a prime mover for a wide range of semi-trailers of a variety of specialist natures, such as the Queen Mary aircraft transporter.

The Canadian range of 30 cwt GS trucks was in production until 1943 when, in line with British policy, they were phased out in favour of the 3 tonner. However, 4 × 4 versions, which

*The FWD HAR 4 × 4–2 6 ton GS was an American-built tractor unit coupled to a Canadian semi-trailer supplied to the British. The semi-trailer could be converted to a full trailer by fitting a two-wheel dolly.*

outnumbered the 4 × 2, continued to be used to carry specialist bodies.

The Chevrolet 1533 × 2 30 cwt 4 × 2 achieved fame as the 'standard' LRDG vehicle, originally supplied to the British in Egypt on the Model 30424-C chassis with all-steel 4B1 body. It was adopted by the LRDG because of its performance when fitted with sand tyres, and because of its quiet running, robust construction and relative ease of maintenance. The base workshops in Cairo raised the sides of the body by about a foot, welded brackets for sand channels to the body sides, cut a hatch in the forward right-hand side of the body for access to radio compartment and extra can carriers, added sandmat brackets on the front wings, and fitted an engine water condenser, stronger rear springs, sun compass bracket and machine-gun pintles.

The 3 ton 4 × 2 class was the big brother of the 30 cwt, and was the most numerous of the wartime load carriers after the 15 cwts. The technical specifications of these vehicles have been widely covered in a variety of publications, and it is intended, therefore, to concentrate on the adaptions of the standard 3 tonner to meet special requirements.

The Bedford OY, Austin K3, 3 ton 6 × 4, and Canadian modified conventional lorries could be fitted out as stores vehicles. Obviously, the internal fittings depended on the type of stores to be carried but, in most cases, they consisted of a combination of steel-shelved lockable cupboards, counters with lockers underneath, a writing desk and filing boxes. Fewer cupboards,

*Guy Ant 15 cwt 4 × 2 infantry truck, similar to the beast on which the author learned to drive.*

*An early production Morris 15 cwt with standard body showing the sides dropped for use with the special Bren gun mounting which gave a good field of fire but left the gunner rather exposed.*

but strong racking, were needed for an MT stores vehicle carrying axles, springs, etc.; whereas, radio and instrument stores needed more cupboard space. Internal lighting from the vehicle battery was provided, and it was usual to fit a security screen of strong wiremesh over the tubular tilt supports and hinged mesh doors over the tailgate. These vehicles were manned by RASC or RAOC personnel and formed part of the mobile workshop establishment.

As there was a constant demand for batteries and battery charging facilities, armoured formations had 'slave carriers', usually Loyd carriers, fitted with banks of batteries and 'jumper leads' to facilitate starting AFVs in the field. Slave battery lorries were also provided, and these carried supplies of spare batteries as well as charging facilities. Two 12 volt generators were driven from a gearbox PTO, and a heavy copper strip or 'bus-bar' was provided, to which batteries were connected for charging. Covered benches could accommodate fifty 12v or twenty-five 24v batteries. Battery repair kits, acid and distilled water carboys, jumper leads, protective clothing and test gear were provided. The slave batteries could be 'ganged' together to provide the varying voltages required for 6, 12 and 24 volt systems. The new batteries carried as replacements also required refresher charges from time to time, to ensure that they were at maximum efficiency when issued.

A demountable water purification plant was often carried by 3 tonners, this was used in conjunction with the water carts to provide pure drinking water for the troops.

The regimental medical officer's lorry has already been mentioned and this equipment could be installed in 30 cwt and 3 ton 4 × 2, or 3 ton 4 × 4 vehicles.

All-wheel-drive 3 tonners were fitted out for roles where their cross-country performance was essential to the specific role. Two examples are the mobile kitchen lorry and the D-D compressor lorry. The former was usually a Bedford QL, but CMP 3 tonners were also used to provide hot meals for mobile formations. A cabinet, installed across the front of the body, housed a water-tank, hand-pump, sink and external drain, shelves, five lockers and two drawers. The cooker unit occupied the right-hand side and was a standard petrol burner, capable of heating five large containers. A working top on the left side covered

*A pre D-Day line up of Ford WOT-2H 15 cwts in an ordnance park awaiting issue.*

the six insulated compartments in which food was kept warm until served. The tailgate lowered to a horizontal position and formed a service platform with access ladders.

The D-D compressor lorry was an entirely different installation, fitted to the Ford WOT-6. The forward sections of the body sides were usually removed, and a skid-mounted Reavell compressor, driven by a petrol engine, was mounted transversely at the front. Four high-pressure compressed air storage cylinders were mounted in pairs, one pair over each rear wheel arch. Between the wheel arches, cradles were provided, into which the smaller bottles were strapped when being charged with compressed air. The compressed air was fed, via a pressure control manifold, through pipes and valves to the bottles. These bottles were part of the amphibious Sherman D-D screen inflation system.

There were other specialist bodies on the QL chassis of which the QLT troop carrier is probably the best known. A long body was designed to accommodate 30 troops plus a driver, the chassis frame was extended, the spare wheel slung under the rear, and two 16 gallon fuel tanks were fitted, one either side of the chassis, in lieu of the standard single 28 gallon type. If used as a load carrier, the side seats could be folded back and the centre row could be removed and stowed under the body. The load had to be carried forward of the rear axle, as any excess weight on the 6 ft extension caused the back end to settle, adversely affecting the steering, it was this characteristic that earned the QLT the nickname 'Drooper'.

The standard range of Canadian 3 tonners started their production life with wooden bodies, but were quickly superseded by various all-steel patterns. Many of these bodies were designed to be easily dismantled and reassembled, and were known as Completely Knocked Down (CKD) types. CKD enabled vehicles to be packed in smaller crates for shipping overseas and saved valuable shipping space. There were various standards of dismantling, and a rather elaborate code was devised to indicate the contents of a case. For instance, Beta-C indicated that the parts for a complete vehicle were packed in two or more crates, required cab assembly then mounting of cab and body to chassis then fitment of wheels.

Vehicles in the 3 ton class were generally considered to be underpowered and Canadian vehicles without governors were constantly overdriven.

Repair and maintenance of normal control 4 × 2 3 tonners was easy, due to their better accessibility than the forward control 4 × 4s,

*The Morris-Commercial C8/GS 15 cwt 4 × 4 was introduced in late 1944. It could be fitted with either wood or steel GS or light warning radar body. It was noisy, slow and vibrated terribly, at least the one the author drove did.*

*The personnel/utility 8 cwt was produced in the early war years and the Morris Commercial PU/Mk 2 4 × 2 was a typical example. The tilt could be removed and used as a tent, folding ground spikes providing supports.*

*The Humber FWD had IFS and the chassis carried heavy utility (Humber box) light ambulance and light recce car bodies as well as the PU body illustrated.*

*The Canadians also produced limited numbers of 8 cwts and the Chevrolet 4 × 2 with no. 12 cab is a typical example. The legend C–8–1014 is the original type designation.*

*Bedford OYD engine compartment showing the 6 cylinder 72 bhp Bedford engine and the space under the square bonnet originally intended for large capacity air cleaners.*

Right *Most armies used staff cars that were basically civilian types in 'war paint'. This restored Mercedes sports a complicated but effective camouflage scheme.*

*The other type of light lorry in British service was the 4 × 2 30 cwt typified by the Austin K30. Large numbers were lost in 1940 and production tapered off in favour of the 3 ton class.*

*The 4 × 4 load carrier was intended to replace the 4 × 2 in production but this was never achieved. The Bedford QL was perhaps the best known British 4 × 4 3 tonner. Shown here fitted with a US-style steel cargo body.*

*The 6 ton class of British load carriers was basically made up of civilian vehicles with military type bodies and was very reliable. The example shown here is a Dennis 'Max' Mk II.*

*The Leyland Retriever 6 × 4 3 ton chassis was fitted with a wide variety of different bodies – here is an early production 3 ton GS lorry.*

*A late production Albion BY5 6 × 4 3 ton with the all-steel GS body which superseded the wood body in 1944. A hip ring has also been fitted to the cab tilt.*

where engine servicing suffered. British chassis frames stood up well, but one US type suffered persistently from cracked frames at a weak point behind the cab. Springs, spring hangers and front beam axles all gave some trouble on vehicles in this class.

The 6 ton class of load carriers, because they were generally standard commercial diesel-powered lorries with military body and wheels and a good power/weight ratio, gave a better account of themselves. They were produced in small numbers by firms specialising in 'heavies' such as Foden, ERF Dennis and Maudslay, and were employed in lines-of-communication supply columns.

The above remarks can be applied to the British 10 ton GS trucks. On the early wartime types, such as the Leyland Hippo Mk 1 or the Albion CX6N, separate semi-elliptic leaf-springs for the rear axles limited their off-the-road performance, but later designs had articulating bogies, e.g. Foden DG6/12. The Leyland Hippo 2 was a wartime design with an open cab similar to the Mk 1. A closed cab was quickly substituted, as the vehicle was intended for long haul supply routes from the Normandy beaches to forward supply dumps. A thousand Mk 2s were in service by the end of the War. It is worth noting that no 6 × 6 10 ton GS vehicles were produced by British manufacturers until after the War, although a 6 × 6 short wheelbase prototype load carrier had been constructed by Leyland in the mid 1930s. Presumably lend-lease supplies of 6 × 6 vehicles were sufficiently plentiful not to warrant production in Britain. Like the 6 ton class, the 10 tonners were reliable vehicles, even in North Africa the best were able to reach 25,000 to 30,000 miles without major trouble, at which mileage, an engine change was usually required. Their American counterparts were not as reliable under the same conditions, as radiators and electrical systems tended to cause trouble.

The semi-trailer load carrier, sometimes described as articulated, was confined to the armies of Britain, the Dominions and the USA. The advantage of the semi-trailer load carrier, apart from manoeuvrability, was that it could carry twice the load of an equivalent GS tractor unit, e.g. the early Bedford-Scammell OXC was classed as a 3 tonner whereas the Bedford OX 4 × 2 GS WD was a 30 cwt, the weight being distributed evenly between the tractor turntable (fifth-wheel) coupling and the semi-trailer axle. The pre-war Bedford Scammells and Scammell Mechanical Horses had been designed with patent automatic quick-release couplings for use in confined spaces, such as railway marshalling yards. They were usually supplied with two trailers in order to double the utilisation, which was popular with everyone except the driver!

*The Ford F30L was typical of the early CMP 30 cwt body type 3A1 and was discontinued in favour of the 3 ton class. The 3 ton version was similar in appearance but with heavier springs and larger tyres.*

*The standard US $1\frac{1}{2}$ ton $4 \times 4$ cargo truck produced by Chevrolet (Model YP-G-4112) was supplied under lend-lease to Allied Armies.*

Left *Austin 10 hp G/YG light utility. Based on the pre-war Austin civilian chassis, the 'Tilly' was widely used for light general transport duties. Similar designs were produced by Morris, Hillman and Standard.*

The majority of Bedford-Scammell load carriers supplied to WD contracts, had two trailers per tractor, in addition, there was a wide range of special purpose semi-trailers available, particularly for the RAF. Not all load carrying Scammell semi-trailers were interchangeable, due to minor changes in the tractor coupling-gear.

A wide range of tractor/semi-trailer combinations was introduced, to make up the deficiencies in availability of the heavy-load class of conventional lorries. Tractors were available from lighter chassis production lines. A 4×4 3 ton tractor, coupled to a 6 ton semi-trailer, could do the same work as a 6 ton 4 × 2 with better cross-country performance; the rig was also more manoeuvrable than a conventional 3 tonner towing a 3 ton trailer.

Top Right *The Dodge T215 ½ ton 4 × 4 WC23 command reconnaissance vehicle eventually superseded by the ¾ ton version, but many survived in service into the post-war years.*

Bottom Right *Ford of America delivered many chassis to Ford at Dagenham to be fitted with British type GS bodies. The model shown is an E.O.15T.*

Bottom *The Dodge T214 ¾ ton 4 × 4 weapons carrier was the US equivalent of the British 15 cwt platoon truck and small numbers saw service with the British Army chiefly in India and the Far East.*

PASS

Some of these combinations were permanently coupled with a ball-type swivel, examples being the CMP Ford F60T 6 ton, 4 × 4–2, GS and the Bedford QLC 6 ton 4 × 4–2, this tractor having a small GS body of 1 ton capacity. The FWD COE and HAR-1 were also supplied as tractors, with 10 ton and 6 ton GS trailers respectively, but these could be detached and a dolly was supplied to convert the semi-trailer into a full trailer. US semi-trailers had a standardised coupling, but interchanging between tractors was not always feasible, for example, the Studebaker $2\frac{1}{2}$ ton 6 × 4 tractor had vacuum brake connections while the International KR-8 4 × 2 had air pressure brakes.

Semi-trailers of all kinds were widely used in the supply columns of the NW Europe theatre of operations, but earlier in the War, Bedford-Scammells formed the large part of the transport element of the British force sent to Norway, as it was thought their manoeuvrability would be an asset on the dockside during disembarkation and on the winding Norwegian roads.

Despite the great degree of standardisation of cargo vehicles in the $1\frac{1}{2}$, $2\frac{1}{2}$, 6 and 10 ton classes, the Americans still had a vast complement of limited standard and non-tactical or administrative vehicles for domestic use or employment on bases overseas. The cargo bodies were similar in design, varying only in size. Of all-steel construction with low fixed-sides, the cargo bodies were surmounted by slatted 'stake sides', the lower four slats of which formed a folding troop-seat while the upper two slats formed a fixed back rest. The canvas tilt was supported on transverse bows, the number varying with the body length. There were no fore and aft supports, and this resulted in the typical sagging appearance of US tilts. After mid 1943, most US standardised vehicles were fitted with open cabs and canvas screens, this at a time when British design was going in the other direction. One major advantage enjoyed by the US trucks was that the majority were equipped with a winch that was most useful for self-recovery.

Typical examples of standard US cargo trucks were:

Truck $1\frac{1}{2}$ ton 4 × 4 Cargo (with winch) Chevrolet YP-G-4112

Truck $2\frac{1}{2}$ ton 6 × 6 Cargo (with winch) GMC CCKW-352

*The Studebaker $2\frac{1}{2}$ ton 6 × 6 cargo truck was produced in short and long wheel base versions mainly for lend-lease and influenced Russian post-war design.*

Truck $2\frac{1}{2}$ ton 6 × 6 Cargo 15 ft Body COE GMC AFKWX-353

Truck 6 ton 6 × 6 Cargo (with winch) Corbitt 50SD6

Truck 10 ton 6 × 4 Cargo Mack NR9

The vast quantity of trucks produced has meant that many have become available for preservation by private enthusiasts, while some are still serving in armies of the world to this day.

Like the other main combatants, the Germans had to decrease their level of standardisation, by including commercial or converted civil designs, to achieve the quantities required by the military.

Many of the designs were complicated and were dropped in the interests of economy and greater production. One of the victims was the

Top Right *Morris CS8 truck 15 cwt 4 × 2 GS. The absence of tilt and driver's canopy reveal the very basic layout of this 'Infantry Truck' with no concessions to crew comfort.*

Bottom Right *Chevrolet YP-G-4112 truck $1\frac{1}{2}$ ton 4 × 4 cargo. Produced in large numbers from 1940 the basic chassis was fitted with many different specialist bodies besides the cargo body illustrated.*

Z536407

USA
3479286
1A ORD
P26

*The Hillman 10 hp light utility Mk 1 was of the class of 5 cwt trucks based on saloon cars of the period and known as 'Tillies'.*

*This interior view of an Austin 10 hp G/YG light utility shows the folding rear seats, tilt hoops stowed and various stowage lockers.*

Klöchner-Deutz 6 × 6 light lorry. All six wheels were suspended on long wishbones, the rear bogie being balanced by bellcranks. Differentials were bolted to the chassis frame and each wheel was driven by a propeller shaft. There were, however, good simple designs available in the 4 × 2 $4\frac{1}{2}$ ton range, many of which were produced as 4 × 4s with 110 bhp diesel engines. In line with European practice, these trucks were capable of towing trailers up to a gross weight of 10 tons. A typical example in this class was the MAN ML 4500 S produced from 1941 to 1946.

In view of the elimination of the more complicated vehicles during the Schell programme, it is somewhat surprising to find the Tatra Company introducing a sophisticated heavy cargo truck in 1943. The Types 6500/111 and 8 00/111 had a chassis consisting of a massive

The Humber FWD 4 × 4 heavy utility, commonly known as the 'Box' was the only British four-wheel drive vehicle in its class and was deservedly popular, being rugged and reliable.

tubular backbone to which the three swinging-axles were attached. The drive shaft from the engine passed down the centre of the backbone, and axle drive shafts were of an ingenious if complicated design that eliminated the use of universal joints. One problematic feature of the design was the use of small diameter tyres on the twin rear wheels, this led to all of the load being borne by only one tyre of each pair during spring deflection.

The engine fitted to this vehicle was equally interesting. It was a 14,825cc V-12 diesel, developing 220 bhp at 2,250 rpm, air-cooled by two 16 bladed ducted axial fans, driven by shafts from the front cover. This vehicle continued in production, in modified form, for several years after the war.

*After 1940 undelivered vehicles ordered by the French from the US were diverted to Britain. Among them were quantities of GMC Model ACK-353 $1\frac{1}{2}$ ton 4 × 4 cargo trucks.*

# OPERATIONS

*A supply convoy of British Army Dodge T-110L 3 tonners moves up through a Normandy village.*

Various types of specialist and load carrying vehicles have been described, and a brief account of the operations that took place in various British theatres of war should serve to give the reader an idea of the problems involved in maintaining fighting forces in the field.

It should be realised that a division and its supporting troops require a daily delivery of approximately 520 tons of supplies. These must be positioned at the division's forward maintenance area for collection by unit transport. A 180 tons being required daily to support the lines-of-communication troops. These supplies can be divided into three categories – food, petrol and ammunition. The amount of food consumed is fairly constant, while the ratio of petrol to ammunition required varies inversely, being dependent on whether the division is advancing or in a defensive position.

A general transport company of the RASC was attached to each brigade of a division, plus additional divisional troops companies. The GT company consisted of an HQ composite platoon for accounting, administration, etc., and three transport platoons plus a workshop platoon. Each transport platoon, commanded by a lieutenant had 30 vehicles on charge plus three in reserve, all 3 tonners, together with six light reconnaissance cars for convoy protection, a kitchen lorry, a Jeep and seven motor-cycles. The five sections of each platoon were led by a corporal in charge of six 3 tonners.

In 1940, the BEF had 104 transport units in service and totalled 21,000 vehicles. Unfortunately, the majority of these were lost in the

*A group of Eighth Army 3 tonners laden with jerricans prepare to move off. All are Canadian built and the nearest vehicle sports a sun compass bracket on the roof.*

evacuation. It is worthy of note that, in 120 days, increased production made good the total losses of wheeled vehicles in France. Many of the stores, particularly tinned petrol, had been prepositioned in dumps prior to the outbreak of the war in the BEF zone of operations, but changes in the method of distribution were made as a result of this early campaign.

In the Western Desert conditions were totally different from those in France and the Low Countries, and it has often been said that the desert was a tacticians dream and a logisticians nightmare. The distances were enormous, the going rough to terrible, and relatively simple problems such as signposting or locating units became aggravated, as when local Arabs removed signposts, prizing the steel poles and using the signs as firewood. 3 tonners were downrated to a maximum of $2\frac{1}{2}$ tons, and average speeds over rough going rarely exceeded 5 mph, while bogging down in soft sand, particularly with 10 tonners, was all too frequent. On top of all the other problems, was the problem of providing sufficient water. The average issue was 1 gallon per man per day, and the water supplies required at least 30 more 3 tonners than did petrol, even though petrol packed in 4 gallon tins lost 30% of its original contents due to evaporation. Unfortunately bulk carriers were not available in sufficient quantities.

In 1939, the Mobile Force, forerunner of the 7th Armoured Division, had a transport fleet of only 96 vehicles; but in 1940, deliveries from Canada and the US arrived direct from North America, and included 4,000 Chevrolets. Many of these arrived in CKD form and assembly plants had to be set up. Four assembly plants in Egypt turned out 180 per day, totalling 45,000 vehicles by October 1942.

Prior to the Battle of Alamein, the ebb and flow of the campaign had taken its toll of vehicles, and there were only just enough vehicles to support the advance of the Eighth Army.

The revitalised Eighth Army had 36 GT companies, nine water companies, one bulk petrol company and six tank transporter companies (one of which was equipped with 10 ton GS lorries, as transporters were not available) providing its third line transport. By the time Tripoli was taken, the nearest railhead was 1,000 miles away at Tobruk and the GT 3 ton companies were now raised to 58, with an additional seven 10 tonner platoons; these were just about capable of maintaining the Army's requirements.

*A Divisional HQ signals lorry set up for field operations. A sign reads 'public telephone' and the vehicle is an early CMP 3 ton GS.*

Because of the long distances to be covered without the provision of lines-of-communication depots, extra rations and fuel had to be provided to enable convoys to make the return journey. Obviously, it was not much use delivering 50,000 gallons of petrol to a forward supply base and then requesting an issue of 8,000 gallons for the return trip, as would have been necessary for a 200 strong supply convoy of 3 tonners running between Tobruk and Msus, so up to 40 additional lorries were required for convoy self support.

Much use was made of captured enemy transport, particularly in the early days of desert warfare, and the Italian diesel powered heavy trucks were highly regarded, although many were put out of action by enthusiastic British drivers before they could be assembled into homogeneous transport units.

The invasion of North Africa by the combined British and US forces brought with it a series of new supply and transport problems. However, although the enemy were still capable of putting up fierce resistance, the Allied forces had the advantage of harbour facilities for immediate unloading.

*The Chevrolet NG-G-7105 1½ ton 4 × 4 panel truck was a rare vehicle in British service, it was designated K-51 by the US Signal Corps. Shown here in Italy with the Eighth Army.*

One of the first tasks to be undertaken was the delivery of 100-octane petrol to airfields, for use by Allied fighters flown in from Gibraltar. This fuel had been prepacked in drums and loaded into tippers which drove straight from the landing craft to the airfields, to be followed by a 6 ton platoon in the next wave.

There was a grave shortage of transport, due to several factors; ships containing essential vehicles had been lost *en route*; the French forces had no supply vehicles of their own, and the American supply vehicles had not yet arrived – which meant transport designated to support two brigades was having to cope with double the task; the weather was atrocious and the ships carrying the vehicle workshops took a while to dock. Many of the British heavy load carriers had, in fact, been requisitioned and were of varied types, thus adding to the problems of maintenance and provision of spares.

The Allied First Army had a strength of a quarter of a million men, but had only ten GT companies, four troop-carrying companies and three tank transporter companies plus tipper and bridging units; less than the Eighth Army.

However, despite the difficulties and shortages the campaign was fought to a successful conclusion. Some idea of the transport task can be gauged from the following statistics, during the seven months of the Tunisian campaign, over 35 million miles were logged by just over 6,000 load carriers, and one particular bulk fuel-tanker logged 30,000 miles – the average was 15,000 miles per vehicle. One unit transported an American combat team 497 miles in 41 hours, while the tank transporters were in constant use, and suffered heavily from enemy attack (16 in one air strike) and from mechanical attrition.

There were two week 1,500 mile round trips from Algiers to Oram to Bene to Algiers or 'short' deliveries averaging 165 miles, one company transported over 1,500 AFVs while 65 Diamond 'T's travelled over 600,000 miles or 9,850 miles each, and only two had to be withdrawn from service to base workshops, the remainder being kept servicable by the unit mobile workshops.

The clearance of North Africa led to the preparations for the invasion of Sicily. This operation proved the value of the DUKW, and 230 supported the Eighth Army assault force with three platoons attached to 1st Canadian Division.

The Italian theatre of operations presented further transport problems and changes in organisation had to be made as a result. The mountainous back-bone of Italy, slotted by fast rivers, was very different from the vast open desert from which the Eighth Army had emerged. One immediate result was that more bridging companies were established. There were times when a Bailey bridge was erected every mile, and on one occasion nine were put up within the space of 3 miles.

The rate at which engineering stores were consumed, and therefore had to be transported, can be gauged from the fact that, on average,

*A line-up of lend-lease Studebakers after being handed over to the Russians following delivery across Persia from the Gulf.*

800 tons of bridging and 400 tons of general engineer stores were issued daily in the winter of 1944.

The supply and transport problems presented by mountain ranges, were largely solved by the formation of Jeep and trailer platoons and mule companies, while in extreme conditions, porters were used. The 3 tonners would bring supplies as far up as possible, then the Jeeps and trailers would take over and, when fitted with chains, could cope with most conditions, the mules would then pack the stores for the final stretch.

Ammunition resupply was always a problem, added to which gunners managed to position their guns in the most inaccessible places. The Jeep platoons were capable of a total lift of 30 tons, but this pales into insignificance, when one considers that the US II Corps artillery and the British X Corps used over 150,000 rounds, weighing well over 4,000 tons in two days. On another occasion, a Jeep platoon of 28 vehicles delivered a daily average of 11,000 rounds of 25 pdr ammunition for a period of three weeks, moving only by night.

As the advance continued up the Italian peninsula, supply routes were maintained along the coast by DUKWs, and the beach head at Anzio was kept supplied by the novel expedient of taking an LCT of laden 3 tonners on to the beach and withdrawing the previous days now

*These British Army Jeeps in Normandy have stretcher brackets fitted. The fitment of sockets to take these brackets was part of the modification programme carried out in Britain on receipt of all Jeeps from the USA.*

empty 3 tonners which were then returned to base, where they were reloaded for the next day's milk-run.

The terrain in Northern Italy changed from mountains to plains and swamps, and the methods of transport changed with it. We have already discussed how the first LVT operations in Europe were carried out with great success, enabling the Eighth Army to cross the flooded areas of the Po Valley in the closing stages of the campaign.

Apart from maintaining the Allied armies, the transport columns also had to undertake the delivery of essential supplies to the civilian population until local government could assume responsibility. This additional burden was fully appreciated, and suitable arrangements for the task were provided in the Overlord planning.

Before leaving the Italian theatre of operations, mention must be made of the part played by the base workshops set up in captured industrial premises during the advance up Italy. The Breda and Innocenti factories in Rome, a cotton factory and rolling mills in Naples, all became vast workshops, as did the Ansaldo and Caproni works later in the campaign, while Monza became a vehicle park. The whole of this project was staffed by 8,700 British troops and nearly 25,000 civilians, manufacturing vehicle spares and carrying out complete engine and chassis overhauls. One workshop at the Caproni plant in Taleido, turned out 450 vehicles a month, while the firm's engine facility at Saronno overhauled an average of 1,000 engines monthly. Many of the experienced divisions were withdrawn from Italy in preparation for Operation Overlord, leaving their battle worn equipment for the relieving units, who complained bitterly, even though it was pointed out that many of the vehicles had motored from Egypt to the Sangro without a hitch.

The vast quantities of equipment assembled in England, including 37,000 vehicles, started to pour across the Channel once the beach heads were established, and the assault divisions received their full establishment of transport.

*A group of Canadian-built Dodge D15s bearing British census numbers with US 1 ton trailers and manned by Free French about to leave a US beach head in Normandy with supplies for Paris.*

Once again, the DUKWs proved their worth, this time supplemented by the Terrapin. An innovation on the beaches, was the use of Jeep platoons to assemble supply dumps. They collected laden 10 cwt trailers that had been towed ashore by other trucks from landing craft and unhitched at pre-determined assembly points on the beaches.

A great deal of movement was carried out in the confines of the beach heads before the break out was achieved, and the 'bocage' country of Normandy was a traffic controller's nightmare. Very large quantities of ammunition had to be moved, and the 7th Armoured Divisional artillery was consuming 62 3 tonner loads daily. It was at this stage that British infantry first went into battle in improvised armoured troop carriers. Lt Gen Simmonds, GOC Canadian II Corps, ordered the conversion of a number of Priest SP guns into troop-carriers, and withdrew half-tracks and scout cars from engineer and artillery units, in preparation for Operation Totalise. The conversion task was code named Operation Kangaroo and the name became the generic title for this type of improvised armoured personnel carrier.

Once the breakout was achieved, the Allied armoured divisions began their pursuit of the retreating German army. This rapid advance placed an enormous strain on the transport and supply organisations, as only the port of St Malo was operational, and the British 21st Army Group moved their front 300 miles in a week, thus doubling the transport originally planned for, to move the 8,500 tons of supplies needed daily. Such an emergency had been foreseen by the planning staff, and 18 extra transport units were quickly mobilised from home based Anti-Aircraft Command and driver training units. Also, tank transport trailers had sides of Somerfield tracking welded on, and were pressed into service, each one carrying up to 20 tons of stores. Improvised airlifts were also organised, and 4,500 tons of supplies were flown in weekly, until the aircraft were withdrawn for the Arnhem drop. The feeding of the liberated civilian

*A US Army GMC 2½ tonner firmly stuck in the awful conditions of the winter of 1944 in NW Europe. The driver seems to be completely unperturbed.*

population was an additional burden. Paris, liberated over two months ahead of schedule, required a minimum of 2,400 tons of supplies daily, and two columns of 500 civilian trucks plus drivers were dispatched from Britain to assist in this task. The supply situation became serious, and some British and American formations had to be grounded to release the petrol and transport needed to maintain some momentum in the advance. Antwerp was still not operating as a port, and the famous long range convoy routes were organised. These were known by the route signs, and the most famous was probably the Red Ball express. This commenced operations in late August, averaging a daily delivery of nearly 7,000 tons in 1,500 trucks on a return trip of 700 miles, they moved 75,000 tons in 12 days. The Red Lion route delivered 500 tons daily to the British Second Army, to supplement its own transport, and the White Ball ran in support of other American units.

As though there were not enough problems, an epidemic of piston failures rendered over 1,400 British made lorries unserviceable, brought about, it is believed, by the introduction of 80 octane fuel.

The operations in the flooded Low Countries, and the clearance of the Reichswald, saw the introduction of LVTs into resupply operations; and wheeled transport units were quickly converted to the amphibious role, re-equipping with DUKWs and Terrapins, these units being double manned to allow round-the-clock operations.

The build up to the Rhine crossing and Operation Plunder continued apace, and was assisted by the provision of a railway within 3 miles of the front. Despite this, however, 343,000 tons were moved by road transport in a month and a further 152,000 tons in 16 days. Tank transport companies were also hard at work and completed 4,000 deliveries to the Second British Army in the week before the assault. At that time, 21st Army Group had 80 general transport companies using 3 tonners on lines-of-communication duties, with another 20 on construction work, while a further 32 were supporting Canadian First and British Second Armies. Once across the Rhine, it took the British Second Army four weeks to reach the Elbe; not a spectacular advance but still there were logistical problems. The removal of enemy prisoners, the repatriation of liberated prisoners and refugees and the clearance of concentration camps added

to the problems of normal supply over extended lines of communication. At this stage, over 7,500 gallons of petrol were being consumed weekly and two million troops had to be fed and clothed, etc.

It is difficult to convey the sheer magnitude of stores moved by the load carriers of the Allied armies in NW Europe, and statistics can become boring; however, as an example, bulk shipments of fuel to 21 Army Group alone totalled over 1,200 million gallons, together with $2\frac{1}{2}$ million jerricans and $4\frac{1}{2}$ million 4 gallon tins, comparative amounts were required by the American Armies.

The maintenance of the vehicles necessary to move these supplies were also a stupendous task, and workshops were established in liberated factories manned by local civilian labour wherever possible. A total of 176,000 'B' vehicles passed through REME workshops in 21st Army Group between June 1944 and May 1945.

*A 'Red Ball Express' convoy starts out after a brief refuelling stop to deliver supplies to the Allied Armies advancing across Europe.*

*A Bedford MW 15 cwt fitted with an AFV crew-training box in post-war service. These cabins were used to initiate trainees in the mysteries of R/T procedures before they were let loose on real AFVs.*

After the fall of Singapore, a great deal of reorganisation and re-equipment of the ground forces was undertaken and new tactics developed for the reconquest of the lost territories.

Roads were virtually non-existent and the problems of moving heavy equipment through jungles and over mountain ranges were immense. The 14th Army stood and fought even when units were cut off. Delivery and resupply by air became commonplace, even during the monsoon period, and by early 1945 nine tenths of the armies' supplies were delivered by air, including a quarter of a million gallons of fuel during the armoured thrust on Meiktila. However, despite the high volume of air support there was a considerable amount of transport activity needed on the ground. Some of the general transport companies had 3 tonners withdrawn and replaced by 15 cwts and Jeeps as these were considered to be more suitable for jungle tracks. There were also many 30 cwt 6 × 4 lorries still in use, and other vintage transport, together with mules and porters. DUKWs, as already described, played an important part in supplying units near the coast or on navigable rivers, and assisted in river crossing by acting as pontoon tugs.

Perhaps the most spectacular feats were performed by the tank transporter units, of which there were four companies in the area, their main load being M3 Lee/Grants or M4 Shermans.

Prior to the offensive, these units were used to move up replacement AFVs, railway locomotives and river craft; then came the first of many gruelling journeys. On one of them a delivery had to be made to Kohima up the Maniour 'road' and a trial run by a single loaded Diamond 'T' proved it was possible, but only just, as it took over two hours to make the last 5 or 6 miles. Some months later an even more ambitious trip was planned, to deliver a tank squadron to Tamu. The journey was scheduled to take three days. The first day was reasonably uneventful but on the second day there was a long 10 mile grind up a twisting mountain road. The first 5 miles were negotiated in 22, 19, 15, 14 and 12 minutes, the next 3 miles took over $3\frac{1}{2}$ hours to climb. The third day was spent in covering 5 miles of extremely steep track, with hairpin bends that required much reversing to get the transporters round, and two tractors had to be coupled together to get each loaded trailer up the disintegrating road. Thereafter, the route was mostly downhill, even so, there were some short sharp slippery gradients, but these were overcome by the simple expedient of closing up the convoy so each vehicle pushed the one in front, the rear 'banking engine' being the unit recovery tractor. The delivery was finally completed in four and a half days.

Other tank transporter companies supported the 254th and 255th Indian Tank Brigades in the drive on Rangoon and Meiktila, where the 400 mile journey took two months to accomplish, where dust was often a yard deep and where tanks were offloaded 17 times in 20 miles to tow the transporters out of difficulties. The final 350 mile thrust to Rangoon was over relatively flat country, more suited to mechanised warfare, but

*A US Army Engineer column, of Corbitt 50 SD6 prime movers and trailers, moves through the ruins of Munchen-Gladbach.*

MÜNCHNER

*British Army Mack NR9 10 tonner loading supplies at Antwerp. The opening of this port eased the strain on the extended supply lines stretching back to the Normandy beaches.*

was a race against the monsoon, a race that was just won by IV and XV Corps supplied mainly from the air.

The problems of providing suitable local carrying vehicles for operations in mountainous jungle conditions, such as Burma, Malaya and New Guinea, were the subject of much research before the war ended suddenly with the Japanese surrender. A new generation of Jungle 'Buggies' were under development, and several prototypes had already undergone extensive testing, the aim being to provide light, air-portable all-wheel-drive vehicles with some amphibious potential, such as the Standard JAB 4 × 4 with floatable trailer, into which the vehicle could be loaded and ferried across calm water, or carried by four men.

The cessation of hostilities did not bring an end to the transportation requirements of the Allied Armies who assumed the role of occupying powers with responsibilities for rebuilding war-shattered industries, feeding the civilian populace and repatriating prisoners of war.

The lessons learned during the years of conflict were studied and applied to a new generation of prototype vehicles destined to re-equip the armies of the world. Russia had a vast new automotive industry whose military products bore unmistakable traces of their lend-lease ancestry. America strove to maintain her new supremacy in military achievement, while in war-weary Britain, the signs of financial restraint were becoming evident once again.

# INDEX

Page numbers in *italic* refer to illustrations